James Lawrie

The Roman or Turkish Bath

Together With Barege, Medicated, Galvanic and Hydropathic Baths

James Lawrie

The Roman or Turkish Bath
Together With Barege, Medicated, Galvanic and Hydropathic Baths

ISBN/EAN: 9783337021139

Printed in Europe, USA, Canada, Australia, Japan

Cover: Foto ©berggeist007 / pixelio.de

More available books at **www.hansebooks.com**

ROMAN

OR

TURKISH BATH:

TOGETHER WITH

BAREGE, MEDICATED, GALVANIC, AND
HYDROPATHIC BATHS.

By JAMES LAWRIE,
M.D., L.R.C.S.E.

EDINBURGH:
MACLACHLAN AND STEWART.
LONDON: SIMPKIN, MARSHALL, & CO.
MDCCCLXIV.

PRINTED BY NEILL AND COMPANY, EDINBURGH.

TO

SIR JOHN DON WAUCHOPE, BART.,

OF EDMONSTONÉ,

IN ACKNOWLEDGMENT OF

MUCH KINDNESS RECEIVED,

This Volume

IS,

BY PERMISSION, RESPECTFULLY DEDICATED

BY

THE AUTHOR.

CONTENTS.

CHAPTER I.

INTRODUCTORY.

                                                                                     PAGE

Revival of the Roman Bath.—Opposition met with.—Useful discoveries frequently opposed at first, and their discoverers denounced.—Harvey, Ambrose Paré, Jenner, Professor Simpson, David Urquhart, Esq., ... 1

CHAPTER II.

HISTORY OF THE BATH.

Origin lost in Antiquity.—Earliest discovered remains in Baalbeck, Phœnicia.—Hamâm Ali at Baghdad.—Among the Ancient Persians, Egyptians, Lycians, and Greeks.—Magnificence of the Roman Bath.—Why called also the Turkish Bath.—Remains of Roman Baths in England.—Numerous and interesting Remains in Scotland, particularly at Inveresk, Fisherrow, Cramond, Croy, Castlecarry, Duntocher, Carstairs, and Burghead.—Its reintroduction into this country in Modern times, ... 12

CHAPTER III.

NATURE OF THE BATH.

Different from all other Baths.—Heated Air its essential principle.—Mode of Construction.—Description of the process.—Vestiarium—Tepidarium—Sudatorium—Shampooing—Lavatorium.—Delightful Sensations.—Frigidarium.—Points of agreement and difference between the Modern Turkish Bath and the Ancient Roman Bath, ... 43

CHAPTER IV.

THE BATH IN DIFFERENT COUNTRIES.

PAGE

Very widely spread throughout the World.—Introduced into Ireland by the Phœnicians.—*Tig Allui*, or Modern Irish Sweating-Houses.—Among the Mandingoes—North American Indians—The Chinese—The Russians—In Northern Europe and Central Asia—The Moors—Modern Egyptians—Turks.—Described by Bonar and M'Cheyne, Eliot Warburton, Savary, Lady Mary Wortley Montagu, and N. P. Willis.—Ancient described by Celsus, Pliny, Seneca, Gibbon, and Count Rumford, ... 71

CHAPTER V.

PHYSIOLOGICAL ACTION OF THE BATH.

Anatomical Description of the Skin.—Seven Millions of Pores.—The Skin a Breather.—A Maintainer of Animal Heat.—Its vicarious action.—*Modus operandi* of the Bath.—Heat.—Action of Heated Air on the Body.—The Bath a dose of heat, 102

CHAPTER VI.

USES OF THE BATH.

Cleanliness.—Care of the Body a duty.—The Body God's gift.—Modern neglect of Personal purification.—The Bath a Purifier of the Body.—A Beautifier of the Complexion.—Irish Peasant Girls.—John Bunyan.—Cosmetics and enamelling of Faces.—A Preserver of the Health.—A Fortifier of the Body.—A clean Skin generates its own Heat.—Hippocrates—Islamism and Temperance.—Bacon.—Cases, ... 121

CHAPTER VII.

PROPHYLACTIC AND THERAPEUTIC BENEFITS.

Importance of Prophylactics.—Professor Simpson.—The Bath preventive of Disease.—Dr John Brown.—Therapeutic or Curative in its Action.—Testimony of Dr Goulden and Sir John Fife.—Remarkable Cases.—Dr Brereton, Dr Spencer

Wells, Dr Haughton.—Sir Benjamin Brodie and Baron Alderson.—Professor Henderson.—Dr Cummins, Dr Erasmus Wilson.—Its efficacy in Gout, Sciatica, Rheumatism, Tic, Neuralgia, Skin Diseases, Ague, Dropsy, Dyspepsia, Bronchitis, and Phthisis.—Remarkable Cases.—Insanity.—Dr Yellowlees.—Lunatic Asylum, Cork.—Testimony of Dr Power, 144

CHAPTER VIII.

OBJECTIONS ANSWERED.

Supposed danger of catching Cold.—Absurdity of.—Illustrative Cases.—Weakening Effect.—Falseness of this notion demonstrated.—Fear of Determination of Blood to the Head.—Editor of Glasgow Medical Journal disposed of.—Dr Lawson of Birmingham ditto.—Supposed Danger from high degree of Heat.—Chimerical Fear.—Violent and Unnatural, and a Novelty in this Country.—Foregoing Objections satisfactorily disposed of.—Facts showing the folly of the Glasgow Editor.—Minor Objections.—Convincing Facts, ... 183

CHAPTER IX.

GENERAL CONCLUDING OBSERVATIONS.

Hydropathic Treatment of Disease.—Striking Cases.—Absurd Prejudices.—Expensiveness considered.—Turkish Bath Treatment as applied to Cattle.—Favourable Notices of Sciennes Hill Turkish Baths.—A Visit to.—Dr Alexander Wood.—Complimentary Testimonials.—Turkish Baths in Sydney, . 234

CHAPTER X.

BAREGE, MEDICATED, AND ELECTRO-GALVANIC BATHS.

Antiquity of the Barege Baths.—The Nitro-Muriatic Acid Bath.—The Electro-Chemical Bath.—Numerous Cases.—Inhalation of Oxygen Gas, ... 270

LIST OF ILLUSTRATIONS.

	PAGE
Hypocaust at Inveresk,	28
Hypocaust at Duntocher,	31
Sculpture found at Nether Croy, representing a Female coming out of the Bath,	33
Ground Plan of the Roman Bath, Castlecarry.	34

THE TURKISH BATH.

CHAPTER I.

INTRODUCTORY.

The Roman Bath bids fair to become once more a domestic institution in Great Britain. Its reintroduction into this country, and the benefits which it is fitted to confer upon all, are at the present time beginning to excite a large share of public attention. My object in this work, therefore, is to present to the reader a clear and intelligible statement of facts respecting the history, nature, and uses of the Roman or Turkish Bath, and the great benefits to be derived from it, not only in the preservation of health, but also in the cure of various diseases. I could have wished that some one better qualified than I am should have taken the initiative in presenting and urging the claims of this powerful prophylactic and therapeutic agent upon the attention of the community; and, were it not that I have a firm and deep conviction of the great value of the Roman Bath as an excellent remedy for many painful diseases,

I would not have ventured to obtrude my views upon the public. But having been for a considerable time past engaged in the administration of hydropathic appliances, and various kinds of medicated baths, together with the Roman or Turkish Bath, and finding that the general public—and even many of my medical brethren—are either ill-informed or apathetic and indifferent with regard to the last of these, I feel constrained to come forward, however reluctantly, to explain what a Turkish Bath really is, and what it is fitted to accomplish when judiciously administered.

It is well known that there is the utmost difficulty in inducing mankind to judge of things upon their own intrinsic merits. They will not undergo the labour and trouble of examining matters for themselves, but on all occasions prefer going on as their fathers did before them, and have no tolerance for those who endeavour to lead them from the beaten track to which they have been accustomed. In their eyes, it is enough to condemn the most valuable discovery, that the sin of novelty attaches to it. Hence the apostle of reformation often dies as a martyr to it, and seals with his blood those truths that are fraught with blessings to future generations.

The Roman Bath is generally esteemed a novelty in this country, although the real truth is, that what is now sought is simply the revival and restoration of that which did at one time exist and flourish as an established institution in these realms. But, true to the history of all novelties in which precious truth is wrapped up and contained, it has had to fight its way onward to public notice, through much violent opposition, and in spite of many apparently insurmountable difficulties. "Perhaps," says Dr Rayner, " there are few things more remarkable in the history of the world than the tenacity with which men cling to old prejudices, and the hostility with which they combat new doctrines and ideas." Innumerable illustrations of the truth of this remark may be found among all nations, in all departments of science, politics, and religion. Instead of drawing forth instances from other fields, however, let me only refer to a few connected with my own profession—that of medicine itself.

In looking, then, into the annals of discovery, we find that when the circulation of the blood was first made known by the celebrated and immortal Harvey, the announcement of this grand truth was received by the high priests and scribes of the medical profession with incredulity, denial, and abuse, and its distinguished promulgator was actually denounced as an impostor and a quack. Again, when Ambrose Paré, an eminent French surgeon, introduced the humane method of ligature in amputations, in which the hæmorrhage from the divided arteries was arrested by tying them with a small silk thread, his proposal was received by the profession with stern obstinacy, was long rejected by them, and the cruel practice of searing the surface of the wound with caustics and hot irons continued and preferred; the illustrious innovator himself was assailed with every expression that jealousy and obloquy could suggest, and the idea of allowing a man's life literally to "hang upon a thread," was denounced as little short of insanity. In practical medicine, also, how great were the difficulties, now almost forgotten, accompanying the introduction into medical practice of antimony, cinchona bark, and other useful remedies. And the professional and religious prejudices that were aroused a few years ago, respecting the propriety of saving human life in difficult cases of parturition by means of chloroform, present another case in point. Happily, the honoured and distinguished discoverer of the value of etherization in labour, Professor Simpson, has seen the speedy triumph of his humane and benevolent innovation over the obstacles which it at first encountered.

But, perhaps, the most striking instance of opposition to useful improvement and precious truth is to be seen in the reception at first awarded to the glorious discovery of the immortal Jenner. His first attempts to introduce the practice of vaccination, and thereby abolish the loathsome and fatal disease of small-pox, which used annually to carry off many thousands of human beings, were denounced, ridiculed, and most determinedly opposed. The introduction of this inestimable boon, which has saved millions of human lives since its promulgation throughout Europe and other quarters of the globe, was most bitterly denounced and decried, and every method that the malice and ingenuity of his contemporaries could devise, was employed to retard its progress. Learned medical professors and doctors united in calling it "a destructive practice," which "ought to be prohibited by Act of Parliament." Yea, an Anti-Vaccinarian Society was got up " to support the cause of humanity against the cow-pox injuries," and to suppress the "cruel despotic tyranny of forcing cow-pox misery on the innocent babes of the poor;" and it was inveighed against as " a gross violation of religion, morality, law, and humanity." To such a length did professional opponents go, that they called it " a diabolical invention," " a bestial humour," " a cow-pox poison;" and declared that in process of time it would transform the human race into animals of the bovine species. It was affirmed that persons who had been vaccinated " coughed like cows, and bellowed like bulls." It was averred that horns were seen beginning to sprout on the foreheads of children, that one was growing hairy all over, that the face of another was assuming a resemblance to the countenance of an ox, and another was transmuting into the visage of a cow. A sapient member of Parliament declared, that a certain family in the south of Hampshire that had been vaccinated, had, like Nebuchadnezzar, been turned out into the fields to graze; and so completely had their natures been changed, that the males of this unhappy family imitated the bellowing of a bull, and the females the lowing of a cow. Jenner's discovery was pronounced by the members of his own profession as a tempting of God's providence, and therefore a heinous crime! Its abettors were charged with sorcery and atheism. One writer strongly wished that the inventors were all hanged, and he would give his vote for its being done. And so successfully did his adversaries work upon the fears and prejudices of the educated public, that this great benefactor of our race was actually attacked from the pulpit as a monster of presumption and impiety.

But, what a change since then! Jenner's discovery

is now regarded, from its vast, amazing, and unfailing success, as the mightiest triumph which medicine has ever achieved over one of the most loathsome and dreaded of human diseases. Jenner's statue has recently been erected in the metropolis of our empire,—his name is honoured throughout the world,—and the practice which he sought to introduce is now rendered compulsory by the law of England. Thus we find that truth is ever powerful, and will always ultimately prevail. In the preceding instances it triumphed over all opposition, and in some of them in an incredibly short period of time. The circulation of the blood has long been regarded as an established fact; a proposal to revert to the actual cautery in amputations would be received with horror; and, as regards vaccination, myriads have through its beneficent operation been annually rescued from the jaws of a fearful and desolating scourge, and saved from an early and untimely grave. Nor is there the slightest appearance that, since its introduction, mankind have in any way become more akin to cows and oxen.

But what a lesson does the preceding melancholy chapter in the history of the medical profession teach us! Does it not teach lessons of mingled caution and humility, candour, modesty, and love of truth for its own sake? But is it not the fact, that improvements in medicine have still to run the gauntlet of bigotry and intolerance? Professor Simpson, in one of his learned works, speaking of these forms of opposition to new truths, says, "They are the same by which many of the happiest and greatest improvements in our profession have each in turn been assailed at their first promulgation. From time to time in the march of medicine and other allied sciences, some earnest and expanded mind conceives and elaborates a great and novel thought, destined in its practical application to ameliorate the condition and promote the happiness of mankind. But hitherto, almost as often as the human intellect has been thus permitted to obtain a new light, or strike out a new discovery, human prejudices and passions have instantly sprung up to deny its truth or doubt its utility, and thus its first advances are never welcomed as the approach of a friend to humanity and science, but contested and battled as if it were the attack of an enemy. Further, every proposed improvement seems to be met with the same invariable array of objections and arguments. The discovery may be new, but the grounds of opposition to it are not new—they are merely the old forms of doubt, and difficulty, and prejudice, used on former occasions, recalled and reproduced anew. And the moral is obvious—that while minds anxious to promote new and probable inquiries should not be intimidated and deterred from their pursuit by such prejudgments on the part of others, those who are, on the contrary, anxious to suppress them, should not venture to base their opposition upon mere impressions and mere opinions only."*

Entertaining a high admiration for such noble and liberal sentiments, which so entirely harmonise with what I have already said, while presenting myself as an advocate of the Turkish Bath as a valuable and powerful therapeutic agent, I trust that in what I shall

* Simpson's Obstetrical Works, vol. ii. page 519—*On Anæsthesia.*

advance on the subject, its claims to be received in this light will be regarded by my professional brethren and the public generally with candour and kindness. In attempting to introduce it into the midst of us, as a means of cure in many painful diseases, it has already encountered similar difficulties, and met with similar opposition to that already described in other analogous cases. It has been branded by all sorts of nicknames, and, before having received a fair trial, or, indeed, any trial at all, the most fatal results have been prophesied as likely to arise from its use. For instance, it has been denounced as violent and unnatural, weakening and hazardous, the cure worse than the disease, and so on. Those who use it, it is said, will catch cold, have apoplexy, get heart disease, melt away, and be slowly boiled, stewed, roasted alive, or otherwise done for. For a full answer to such rash and unfounded accusations as these, I refer especially to that chapter entitled "Objections Answered,"* where the absurdity and falseness of all such mistaken ideas are completely exposed. But while the history of medicine and of medical appliances furnishes many warnings against hastily and ignorantly prejudging, that remarkable chapter in it to which I have purposely referred on the first reception given to Vaccination, should operate as an especially humiliating warning.

 The circumstances by which I was first led to an acquaintance with the Turkish Bath, and its value as a most useful medical adjuvant, are briefly as follows: In the autumn of 1857, I had the pleasure of accompany-

* *Vide* chap. viii.

ing the late Richard Whytock, Esq., to Ireland, as his medical attendant, he being about to receive treatment at St Anne's Hill, Blarney. On arriving at that magnificent establishment, I had much pleasure in recognising in Dr Barter, an old friend and class-mate at the University of Edinburgh. I had also an opportunity of conversing with the celebrated Mr Urquhart on the hydropathic system in general, and on Turkish Baths in particular. Mr Urquhart complained much of the want of sympathy on the part of the public in reference to the efficacy of this bath, not only as an invaluable agent in the purification of the skin, but also in the cure of many diseases not easily removed by medical treatment. Although the views enunciated by him were at first treated with ridicule and contempt, and considered as the dreams of a mere enthusiast, that gentleman was not to be discouraged by the envenomed shafts of obloquy and reproach which his opponents so unsparingly showered upon his devoted head. Conscious of the soundness of the principles which he had advocated for twenty years, he determinedly carried out his apparently hopeless labours, and persevered in indoctrinating the public mind by writing and lecturing on the subject, under the full conviction that, by the establishment of Oriental Baths in this country, he would yet confer an immense boon upon the teeming populations especially of our large towns and cities, and a most invaluable means of preserving the bodily frame in a sound and healthy condition. His labours were numerous and incessant, and neither trouble nor expense were spared by him, in order to disseminate a knowledge of this Bath as one of the most powerful remedial agents known to man, no less than as an effectual purifier of the skin.

Dr Barter saw at a glance the great value and importance of Mr Urquhart's views, and with his characteristic vigour, courage, and persevering energy, instantly commenced carrying them out. While I was there, Mr Urquhart was superintending the erection of the first Turkish Bath erected in that country, namely, at Dr Barter's Establishment, St Anne's Hill, Blarney. I had many interviews with Mr Urquhart, and carefully examined the principle on which the bath should be constructed. Having thus seen the first Roman Bath established in these kingdoms in modern times, and being deeply convinced, from the subsequent experience which I have acquired, of the great value and importance of this bath, I readily yielded to a requisition which I received from a number of respectable gentlemen in this city to erect such an institution here. Accordingly, in fitting up my establishment, the first of its kind in Edinburgh, the same arrangements and plans have been in general adopted as I saw at St Anne's Hill, and no expense has been spared in rendering it elegant and comfortable, as well as perfectly safe and salutary. And during the past three years, I am happy to say it has been attended by several thousands of ladies and gentlemen from all parts of the country.

To David Urquhart, Esq., late M.P., most undoubtedly belongs the merit of introducing the Roman Bath into this country once more, after long centuries of oblivion and neglect. No one can read his famous work,

"The Pillars of Hercules," especially the chapter on *The Bath*, without a feeling of admiration for his talents, his earnestness, and his learning. For his eloquent and forcible advocacy of its blessings and advantages, and for his enlightened and philanthropic exertions to establish and restore it in Western Europe, he is entitled to occupy a high place among the benefactors of the age.

CHAPTER II.

HISTORY OF THE BATH.

In this chapter I propose to give a brief sketch of the history of the Bath, of whose properties and virtues I design to treat in this volume. After speaking of its history, I shall then proceed to discuss its Nature and Uses—what it really is, and does; and as the Roman or Turkish Bath is entirely singular and altogether peculiar in its nature, which is to be afterwards explained, I request the reader not to confound it, in the meantime, with any other ordinary sort of bathing in the element of water, either hot or cold; for the essential element of this bath, into which the bather is introduced, and from the action of which he derives all its principal benefits, is not water at all, but another element altogether different. The undeniable virtues of water, and its beneficial effects on the functions of the skin, are certainly better understood and more abundantly experienced in this country than in many parts of the Continent of Europe, where the traveller, as in some parts of France for instance, can hardly obtain more water for a bath than he could hold in the hollow of his hand, and where personal ablution is shockingly neglected. And although this was long the case, and is still to a large extent the case with numbers even in our own country, yet the persevering homilies of the medical profession have had the good effect of establishing the water-bath amongst us as a permanent institution. The strengthening and tonic effect of cold water, and our frequent ablutions in that element, both in private and on the sea-shore—while fortifying the constitutions of our people—are, I believe, the wonder, though not the envy, of many continental nations. Warm-water baths, too, have been introduced among us, together with vapour-baths of various kinds. But the bath whose history I am now about to trace is different from, and transcends them all. Our notion of warm baths is merely a great deal of hot water and soap, or perhaps of hot vapour with medicaments: but this is not the hot Bath of the ancient Romans; this is not the Bath prescribed by Galen, Hippocrates and Celsus—the Bath whose praises Homer sings in the Odyssey—which the Greeks taught to the Romans, and of which Julius Cæsar brought the secret even to these islands.

The origin of the Bath is lost in the mists of antiquity. But there is little doubt that it had its origin in the East, where the human race were first planted, and where arts and sciences, knowledge and civilisation, first flourished. Interesting remains of such baths, whose antiquity tradition does not attempt to fathom, have been found in the district of Lebanon, in Syria— the Phœnicia of the ancients. The earliest of these remains have been discovered at Baalbeck, which signifies *The House of Baal*. In these Phœnician ruins, a crypt was found by Mr Urquhart, in which were mortar and charcoal, recognised at once as *kissermil*, or ashes from the bath. The baths had been heated, according to the old construction, in the same manner as an oven, with underground fires. "It was curious," he says, "to observe these ashes in the midst of Cyclopic blocks. And yet why should not the bath have belonged to the very earliest period of human society? It is sufficiently excellent to be from the beginning."* In opening up the pavement of an ancient bath on the western coast of Africa, he came upon a similar deposit in large quantities under the floor. The great antiquity of this bath, and the fact of its existence among the Phœnicians, are farther confirmed by the fact that Homer mentions their baths in the time of the Trojan war. Homer's references to the bath are not unfrequent. He tells us that Minerva discovered certain baths to Hercules, that he might replenish his strength after undergoing severe exertion and fatigue; and he informs us how Andromache, with matronly care, prepared a hot-bath for her husband Hector, against his return from battle.

From Phœnicia, from the coast of Tyre and Sidon, a knowledge of this bath spread along the coasts of the Mediterranean, and notably to Greece, for the Greeks got everything from the Phœnicians, letters as well as baths. The baths of Greece were celebrated for their magnificence, and were usually annexed to the *Palæstra* or *Gymnasia*, of which they were considered a part. There were others, however, not attached to the Palæstra, but for the use of the public, in which men and women were separately accommodated. The Gymnasia, of which

* Urquhart's Lebanon: A History and a Diary.

the Lyceum and the Academia were the most notable and famous, and in the former of which Aristotle lectured on philosophy to his pupils, were spacious and splendid places, in which both mind and body received cultivation, and were exercised for the purpose of producing accomplished scholars and athletic warriors. One of the means for securing this end was the Thermæ, or Baths, attached to every Gymnasium. And these institutions, which at first existed among the Spartans in Lacedæmonia, soon spread from thence to other parts of Greece, and notably to the metropolis of Attica, the famed city of Athens. So that the heads of the republics of Athens and Sparta considered that they had not done their duty, nor rendered the condition of the people tolerable, unless they afforded them the means of enjoying strengthening baths and athletic games.*

But the Phœnician Bath spread into the East as well as into the West. Ainsworth, in his "Journey to Kalah Sherghat," informs us of the existence of such baths at a place call *Hamâm Ali*, or the Baths of Ali, in the neighbourhood of ancient Nineveh, and says that, though commodious only for a half-savage people, yet they are frequented by persons of the better classes both from Baghdad and Mosul. This bath is in principle well known also in Persia, and was in use among other ancient nations of the East, such as the Assyrians and the Medes. Alexander the Great was astonished at the magnificence of the baths of Persia. Indeed, the farther we go into antiquity, the more magnificent

* *Vide* Encyclopædia Britannica—*Baths.*

were the public baths. In ancient Egypt these Baths were also in full operation; and when Alexandria was taken by the Caliph Omar, in A.D. 646, it contained 4000 baths, and the great and precious library of that city, numbering 700,000 volumes, was burned to heat them.

It is curious to find a description of this most effective and singular, as well as most ancient of baths, in the "Arabian Nights' Entertainments." I do not here give that description, as I reserve all account of the principle of the bath to the following chapter. But in that description, the oldest extant of it or of any bath in the world, Abousir, a traveller from Persia, is represented as inquiring his way to the bath on his arrival in Egypt. The Egyptians, in reply, said they bathed in the sea, and asked what he meant by a bath. He answered, "A bath is a place where men are cleansed of the impurities which are upon them, and it is the best of the good things of this world." Abousir, finding that the bath was unknown in Egypt, is then stated to have built one, and invited the king of Egypt to enter it. The king did so; and as the result of the operation he is then said to have experienced a liveliness which in his life he had never known before. Then said the king, "By my head, my city hath not become a city, save by this bath." The queen also entered it, and was so pleased that her bosom became dilated, and she presented Abousir with a thousand pieces of gold; and great honour and good fortune came to him in every way.

From Phœnicia, the bath appears to have penetrated

into Russia also in the north, throughout which immense empire it still exists, and is extensively practised by all classes, though in a modified form. Herodotus says that the Scythians used it at the time of the Trojan war. From Russia, it naturally spread into those Scandinavian countries adjoining, where we find it in operation, in the same rude form as in Russia itself. Dr Clarke, in his "Travels in Europe," says it is common not only "throughout the vast empire of Russia, but also in Finland, Lapland, Sweden, and Norway;" and adds, "while we wonder at their prevalence among the eastern and northern nations, may we not lament that they are so little known in our own country?"

And not only eastward and northward, but into divers western nations also of remote antiquity, did these baths spread, radiating out from Phœnicia as from a common centre. In very early times it was in use all along the southern coast of the Mediterranean, throughout Egypt, Tripoli, Algiers, Morocco, and to the Pillars of Hercules. In the tombs of ancient Lycia, the instruments and processes used in this bath are depicted. The Moors in Mauritania retain it still, and the ruins of ancient baths attest its antiquity among them. It is supposed that by the Moors also they were carried into Spain, for up to the fifteenth century these baths were in common use in Spain, and an ancient law of Castile forbids Moors and Jews to bathe with Christians. But the most remarkable thing of all here is, that Ireland itself received the bath from the Phœnicians; for there was undoubted intercourse between these countries in extremely ancient times. And what is not a little singular, this bath actually did exist among the peasantry of Ireland, although in the rudest possible shape, up to a comparatively recent period, as I shall afterwards show.

But it is now called the Roman, and not the Phœnician, Bath. And why? Because the Romans, after receiving it from the ancient Greeks, carried it to a pitch of perfection and splendour such as it has never attained in any other country in the world. Rome was the glory of these baths, and they in their turn were the glory of Rome. After Greece had been subjugated by the Romans, they carried back with them to Italy the taste for the bath, but the *Balneæ* of the Romans were on a much larger scale than the *Thermæ* of the Greeks, and formed immense establishments,—frequented more for the sake of pleasure and comfort than of health, though it cannot be doubted that they exercised a most salutary influence in promoting the physical well-being of the community. They were public edifices under the immediate inspection of the imperial government. Bathing, as it was then practised, and not as among us—until we know the secret of the Roman Bath, was the favourite and health-giving luxury of the Romans; and the emperors, desirous of securing the good opinion and love of the people, took care to provide them with the means of practising this their chief pleasure, in edifices equal in grandeur and magnificence to the temples of the gods. The most famous of the baths were those of Agrippa, Vespasian, Nero, Aurelian, Domitian, Severus, Maximian, Caracalla, and Diocletian, and the emperors themselves sometimes bathed in public with the rest of the people. The city of Rome possessed at one time no fewer than 850 baths. These buildings were masterpieces of architectural skill and sumptuous decoration. In them was centered all that was most perfect in material, elaborate in workmanship, and elegant in art. They were vast in dimension, elegant in design, and grand in ornament. Ammianus Marcellinus observes that they were built *in modum provinciarum*, as large as provinces. The Baths of Caracalla, for example, were a mile in circumference, and the ruins of this immense structure still remain to this day. They were adorned with 200 marble columns, and were capable of accommodating 2000 persons. But those of Diocletian surpassed all the rest in magnificence and grandeur, and occupied 140,000 men for many years in their construction. From the amazing extent of the existing ruins, the loftiness of the arches, the beauty and stateliness of the pillars, the profusion of foreign marbles, the curious moulding of the vaulted roofs, the prodigious number of spacious apartments, and a thousand other ornaments and conveniences, they form as pleasing a sight to the traveller as any of the other antiquities which remain, and they may justly be regarded as among the most astonishing relics of the luxury and splendour of imperial Rome.

The Romans carried the decoration of their baths to such an unreasonable pitch of extravagance as to call forth reproofs from satirists and philosophers, for the bath was not so much a habit—it was a passion with the Romans. In the reign of Augustus they had reached their highest splendour. Not only were they adorned with the most superb marbles brought from the most distant parts of the world, and with the choicest selections from their rich conquests—the curious and the wonderful in nature and art, together with the finest productions of the painter and the sculptor, such as the Laocoon discovered among the ruins of the Baths of Titus, and the Farnese Hercules in those of Caracalla—but they were decorated even with the precious metals and gems in abundance. The ladies actually had their baths paved with silver. Seneca complains that the baths of the plebeians were filled with silver pumps, and that the freedmen trod on gems. Juvenal lashes, and Fabricius denounces, their excessive luxury; and Pliny wishes good old Fabricius were alive to behold the degeneracy of his posterity, when the very women must have the seats of their baths of solid silver. The wealthy had baths built, frequently of great magnificence, at their own residences. Mæcenas, the Roman knight, and famous patron of learning, has the credit of erecting the first private bath. Statius, in his poem on the Baths of Etruscus, steward to the Emperor Claudius, describes them in the following terms:—

> "Nothing there's vulgar, not the fairest brass
> In all the glittering structure claims a place;
> From silver pipes the happy waters flow,
> In silver cisterns are received below.
>
> See where with noble pride the doubtful stream
> Stands fixed in wonder on the shining brim;
> Surveys its riches, and admires the state;
> Loth to be ravished from the glorious seat."

And Seneca, like a virtuous moralist, sighs for the return of the good old times when baths were places of humbler entertainment :—" But, ye gods ! what pleasure was there in entering those obscure and vulgar baths when prepared under the direction of a Cato, or one of the Cornelii !"

And where, it may be asked, is the Roman Bath at present ? If you look for it in modern Rome, it is not to be found. All this glory has now departed. And however we may attempt to account for the decay and disuse of the bath there, in the troublous times that succeeded the fall of ancient Rome, the fact remains that in those vast halls where the athletæ exercised themselves, and games were played, and poems recited, and orations delivered, and dancing witnessed, and music heard, silence now reigns, unbroken, except by the feet of far-travelled pilgrims from distant lands, come to gaze on the departed glories of the mistress of the world. And he finds that the only practical use to which any part of these immense establishments has been turned is, that the great hall of the Baths of Agrippa is now in all its integrity a place of Christian worship.

Having answered the question why the bath of which I speak has been called the Roman Bath, I now proceed to inquire, Why is it also called the Turkish Bath ? The best reply to this query is to be found in the following eloquent extract from the writings of Mr Urquhart :—" A people who knew neither Latin nor Greek have preserved this great monument of antiquity on the soil of Europe, and present to us, who teach our children only Latin and Greek, this institution in all its Roman grandeur, and its Grecian taste. The bath, when first seen by the Turks, was a practice of their enemies, religious and political; they were themselves the filthiest of mortals; they had even instituted filth by laws, and consecrated it by maxim. Yet no sooner did they see the bath than they adopted it; made it a rule of their society, a necessary adjunct to every settlement; and Princes and Sultans endowed such institutions for the honour of their name."

Thus we find that though the Romans are gone, the Roman Bath is not lost, and that it still lives in its modern offspring the Turkish Hamâm. In the lands of the East, where these baths first rose, they still especially flourish, for, as Disraeli observes, " The East is the land of the bath." In Gautier's " Constantinople of To-day." it is said :—" In the Orient, where cleanliness of person is among the first obligations of religion. the baths have preserved the luxury of both the Grecian and Roman periods, and are superb edifices of the loftiest architectural pretensions, with cupolas, domes, and columns, enriched with marble and alabaster, and displaying great varieties of colour and design, while they are thronged with armies of attendants, whose various offices recall the ancient time already named, and carry us back in imagination to Rome and Byzantium." And in Thornton's " Turkey" we find it stated, that " The Greeks and Romans used the bath as it is still used in the Russian and Turkish Empires, from the northern extremities of Europe to the neighbourhood of the tropics : while the Gothic families, who overspread and settled in the Western Empire, suffered these baths to fall into disuse. The Turks, however, whether they adopted or inherited the custom. found it established in the Eastern Empire, and perpetuated the use of it. The public baths are elegant and noble structures, built with hewn stone; the inner chambers are capacious, and paved with slabs of the rarest and most beautiful marble." Thus we are put in possession of a most natural explanation of the question why the ancient Roman Bath should now only be known to us in the Turkish Bath. Let it only be remembered that the ancient Roman Imperial power lingered out in the eastern part of the empire long after it had been given up in the west: and that at Constantinople the power and authority of the Eastern Emperors continued for many centuries, after Rome and the West had fallen into other hands. The Bath may naturally therefore be sought, and reasonably expected to be found, where the Romans and their power lingered longest, namely, at Byzantium, and in the Eastern Empire, which is the modern Turkey. The invading and conquering Turks found the bath there, and kept it.

But it was from Rome itself that the Roman Bath first found its way into Great Britain. It is known, too, that remains of Roman Baths are extremely common in France, into which country also they must have been introduced by the Romans themselves, as their dominion extended over both Gaul and Britain. Accordingly, we find that extensive ruins of ancient Roman Baths have been found at various places in England, sufficiently distant from each other to show that their use was at one time quite common and general over the country. They have been discovered at Bath, London, Chester, Wroxeter, Hovingham, Cirencester, Colchester, Carisbrooke, Silchester, Arundel, Northleigh, and also at

Wheatley, near Oxford, and at Linley Hall, in Shropshire. The most remarkable and interesting of all these remains are those which have been brought to light at Wroxeter and Chester. The former of these places, Wroxeter, near Shrewsbury, is the ancient Uriconium of the Romans, and was a large and flourishing city. It fell a victim to fire and sword in the fifth century. In the Philosophical Transactions an excellent description has been given of its baths, sudatories, and other remains. Guide-books have been published to its ruins, and a Society has actually been formed for the purpose of excavating its site.

Though Wroxeter is more rich in Roman antiquities in general, yet it must yield the palm to Chester in respect to the subject in hand, as affording the most clear and remarkable evidence of the existence of the Roman Bath in England. The Hypocaustum, or underground heating-place to the Roman Baths discovered at Chester, is, in some respects, the finest and largest that has yet been brought to light in this country. This hypocaust is still in a state of great preservation, and its roof, which was at the same time the floor of the Calidarium, is supported by thirty-two monolith pillars, about three feet in height. Strong square tiles, three inches thick, form its roof, and over these is laid a bed of hard concrete, nine or ten inches thick, the whole being capable of bearing any amount of heat. The floor of the hypocaust has been rendered uneven by the burning embers of its ancient fires; and from the corroded state of the broad, square tiles which compose its roof, together with the large size of this subterranean furnace, some idea may be formed of the prodigious extent and heat of its fires.

But although the remains of the Roman Baths found in many parts of England are large in extent, yet they bear no comparison to the mightier Thermæ of ancient Rome, because here they were pruned of all those accessories which are not essential to the bath as a means of promoting health. Accordingly, all that was merely ornamental or superfluous, all that related to recreation, amusement, and exercise, was omitted, and that which was substantial and wholesome remained. So that the ancient British Thermæ, could we see them reconstructed, would doubtless present us with a model of the bath in its simpler elements, containing only what is required, and all that is required, to constitute the bath, as it has been revived amongst us at the present day.

It is a curious, though little known fact, that remains of Roman baths have also been extensively discovered at various places in Scotland, particularly at Inveresk, Fisherrow, Cramond, Croy, Castlecarry, Duntocher, Carstairs, and Burghead. At Inveresk, near Musselburgh, a few miles to the east of Edinburgh, it is known that a considerable Roman town, gathered round a camp and fortified place, once stood. The entire ridge where the present collection of scattered villas and humble cottages now stands, was occupied by the buildings either of a Roman colony, or Romanised community. On the northern slope overlooking Fisherrow, its harbour and roadstead, immense numbers of ancient foundations of houses have been discovered in turning up and ploughing the fields. And along the whole ridge of the height, from the site of the ancient citadel—which is the site of the present parish church, eastward to Pinkie Burn, so many remains of ancient pavements existed towards the end of the last century, that the plough could not in several places penetrate the ground; and, wherever this was the case, all vegetation was in dry seasons at a stop. The fossé of the Roman citadel, within whose area the church of Inveresk now stands, continued to be visible till within the last sixty years. There is a locality in Musselburgh which still goes by the name of the Camp Close. And the old bridge of Musselburgh, which bears the marks of extreme antiquity, though still very solid and serviceable, is believed, on good grounds, to be the veritable work of the Romans—perhaps the largest and most venerable Roman antiquity which this country can boast of. It is built in the direct line of the old Roman road leading straight down from the citadel on Inveresk Hill to the harbour at Fisherrow. There are bridges standing to this day in France, certainly known to have been the work of the Romans.

More than three hundred years ago, in the year 1547, these remarkable remains at Inveresk had begun to attract attention. In Stuart's "Roman Antiquities of Scotland," it is stated that Queen Mary, in the year 1565, gave orders to the bailies of Musselburgh to " tak diligent heid that the Monument of Grit Antiquitie new fundin be nocht demolisit." The order refers to a Roman altar with an inscription, and to the remains of a Roman bath laid open at the same time a few yards to the east of Inveresk church. And in the State-Paper

Office in London, there has been found a letter from Queen Elizabeth's ambassador, Randolph, to Sir William Cecil, afterwards the famous Lord Burleigh, her great minister, giving an account of the discovery,—" a monument of the Romaynes with words greven upon hym : and Dyvers short pillars sette upright upon the ground, covered with tyle stones, large and thyncke, torning into dyvers angles and certayne places like unto chynes (chimneys?) to avoid smoke." *

There can be no doubt whatever, that this was the hypocaustum of an ancient Roman bath. The writer above referred to, says, " In the year 1783, whilst workmen were engaged in the improvement of some garden ground, a short distance to the eastward of the church, they came, at the depth of 2 or 3 feet, on the floors and foundations of various buildings, which, in course of their operations, were laid open over an extent of 60 feet in length by 23 feet broad. The whole of this space was paved with a kind of mortar, known by the name of *tarras*, and was intersected at intervals by distinct traces of stone walls, among which might be observed the inclosures of two separate chambers, the evident remains of a Roman bath. The largest room measured 15 by 9 feet, the other 9 by $4\frac{1}{2}$. The floors of both were composed of a coating of tarras 2 inches thick, laid, in the case of the first, upon a layer of lime, gravel, and pieces of brick, 5 inches deep, which again rested on a basement of irregular flag-stones,—the whole being supported on rows of pillars 2 feet in height,

* Stuart's *Caledonia Romana:* A Descriptive Account of the Roman Antiquities of Scotland.

some of them formed of stone and others of brick. But in the smallest chamber, the coarse substratum of lime and gravel was 10 inches in thickness, as if this

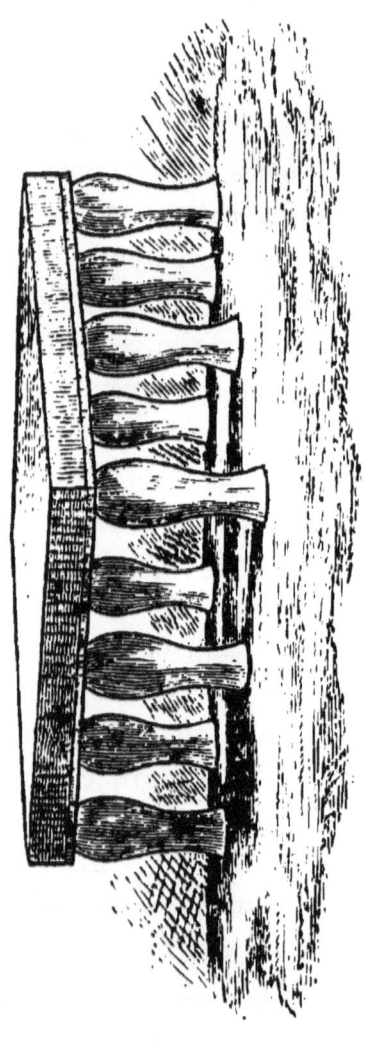

Hypocaustum at Inveresk.

part of the building had been required to sustain a greater degree of heat than the other,—a supposition extremely probable from the circumstance, that the pillars below bore evident marks of having been much injured by fire. A quantity of charcoal was found beside them, in good preservation, as if placed there to renew the glow of a furnace which had been suddenly and for ever extinguished. Under the first apartment, the heat had been conducted by means of flues formed of clay, which were found quite perfect when the discovery was made. The partition wall between the two rooms was pierced near the ground by a hole 3 inches in diameter, through which a pipe of some description had no doubt led as a conduit for water from the one to the other."

To proceed with the account—which was drawn up by Dr Carlyle, the famous minister of Inveresk :—" Such were the most perfect of the ruins brought to light at the period referred to. But all around them were to be seen the remains of other chambers which had evidently been of a similar construction. Taken in the aggregate, they unquestionably marked the position of an establishment of no mean importance in its day— the public baths of the Roman Inveresk. Of their high antiquity there can be no doubt, for every particular mentioned proclaims them Roman. The very cement covering the floors was of a quality unequalled by the skill of later times, and was formed of exactly the same materials as the *tarras* which lined the capacious sewers of 'The Eternal City.' Many other foundations, with a similar description of pavement, were a few years earlier laid open, when a bowling green was being formed to the westward of the *Thermæ* just mentioned. Roman bricks have also been found in considerable numbers in this locality, many of which are said to have been made use of in the walls of the old church of St Michael the Archangel,—a building of great but unknown antiquity, demolished in 1804 to make way for the present structure. Fragments of clay-pipes, common earthenware, and pottery made of a species of fire-clay, have been also from time to time disinterred at Inveresk. The pipes are of an oblong form, fifteen inches in length, with an orifice five inches by three, much discoloured in the inside, evidently from the effects of smoke."

Besides all these remains, a crowd of minor relics were discovered around the dilapidated baths and the before mentioned altar—of Apollo. And at Fisherrow, which appears to have been merely a suburb, and was certainly the sea-port of the ancient and populous Roman city of Inveresk, tradition distinctly reports that the remains of ancient Roman Baths have been likewise discovered, resembling those laid open on the neighbouring height. Ruins similar to this found at Inveresk have been repeatedly exposed in digging the foundations of many houses at Fisherrow.*

At Cramond, the *Alaterva* of the Romans, the remains of baths similar to those met with at Inveresk, together with the remains of subterraneous streets and other traces of mighty masonry, and Roman coins in incredible quantities, have been discovered. Among other varied and curious relics brought from Cramond, and now deposited in the interesting collection at Penicuik House, are to be seen Roman *strigils*, instru-

* See New Statistical Account of Scotland: Edinburghshire.

ments in the shape of a blunt knife, well known to antiquaries, often to be met with in the cabinets of the curious, and equally well known as indispensable implements in the ancient Roman process of the bath.

At Duntocher, the second of the seventeen fortified places built along the line and for the defence of the Great Roman Wall of Antoninus Pius, which stood till at least the year 422 of the Christian era, an interesting discovery was made in 1775. Curious and extensive vaults and subterranean chambers, in which were found some grains of wheat, showing them to have been used as Roman granaries, were brought to light, together with votive altars and other remarkable pieces of sculpture. The Fort and its entrenchments were in 1725 in a tolerably distinct condition, but now scarcely anything of them is to be seen. But, in the year formerly mentioned, 1775, and near the spot where the subterranean chambers were laid open, the extensive remains of a *Sudatorium* were discovered, in one part of which the floor was supported by no less than 144 square pillars of brick, of a beautiful pale red; the

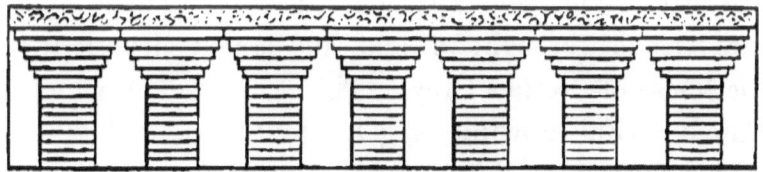

Hypocaustum at Duntocher.

under ones being 8, and the upper ones 21 inches square. On the top of these, as at Inveresk, was spread a layer of lime, mixed with gravel, 5 inches in thickness. Here the garrison of the Fort, and, it may be, the colonists of its vicinity, enjoyed the *deliciæ* of that most universal of Roman luxuries—the Bath.

A very beautiful and remarkable relic of the Bath has been discovered at Croy, a few miles from Glasgow, where stood the tenth of the Roman Forts along the line of Antoninus's wall, built by the Romans to defend their British possessions against the incursions of the fierce Caledonians of the north, who although often attacked were never subdued, and were not unfrequently the attacking party. The hamlet of Upper Croy has been almost entirely built from the Roman ruins, as is significantly shown by the brick-like shape of the stones of which the walls of their houses and even their field dykes are composed. In trenching his garden, a cottager found a number of hollow stone pipes lying in line, and fitted into each other. No great stress may perhaps be laid upon this circumstance, as proving the existence of a Roman Bath there, by those especially who are not acquainted with the construction of a Roman Bath; and neither do I ground much upon it, although the form and position of the pipes are peculiar. But at the same time a curious piece of Roman sculpture was also brought to light, and is now deposited in the house of Nether Croy. This sculpture, of which a very beautiful drawing is given by Stuart, and which, with a few others, I am enabled by kind permission to present to my readers, represents a nude female of elegant figure, with a slight drapery about the limbs, passing through a door-way between two spirally ornamented columns, as if in the act of leaving the bath; while, to her left, another female figure, still more unattired, rests upon one knee in a half-reclining posture, under the fragmentary portion of a sculptured laurel wreath. It is very doubtful if England itself, which was much longer and much more entirely under Roman

Sculpture found at Nether Croy, representing a female coming out of the Bath.

sway than this country, amid all the grand remains of its ancient Roman Baths, of which we have heard so much, has anything to boast of at all equal in interest to what we possess at home, though hitherto almost unknown and unheard of.

At Castlecarry, the twelfth fort, in digging for the formation of the Forth and Clyde Canal in 1769, the workmen came upon the foundations of eight apartments, laid off in the style of a Roman mansion, connected with which were the remains of a " Sudarium, or hot-bath, about 70 feet in length," the ground plan

c

of the whole being distinctly visible. The different chambers of the bath, which, as a whole, must have been admirably and beautifully proportioned, appear as opening the one into the other. Seventy years after this remarkable exposure of the existence of a Roman Bath

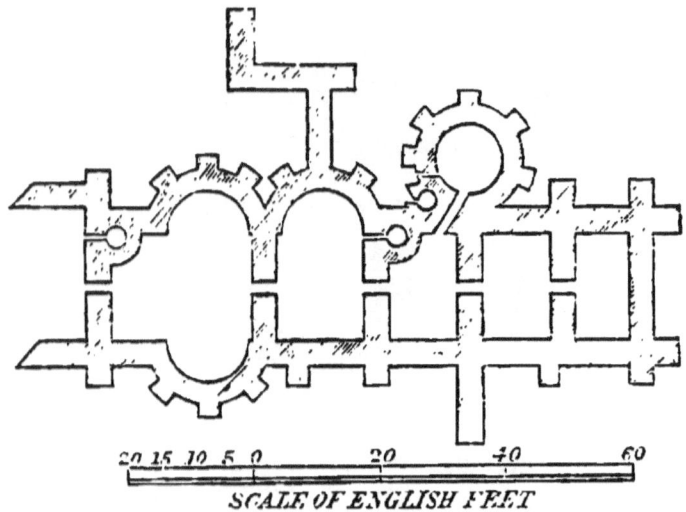

SCALE OF ENGLISH FEET
Ground Plan of the Roman Bath, Castlecarry.

in that locality, the workmen engaged in breaking the ground for the Edinburgh and Glasgow Railway, found that the soil in the same neighbourhood which they had to remove was one entire mass of fragments of pottery, mixed with broken stones, and a multitude of pieces of jars, vases, and basins; these articles having been in far more frequent requisition in Roman Baths than most people are aware of—in fact, from their manner of bathing, in constant use.

In the neighbourhood of the church at Carstairs, near Lanark, the remains of another Roman Bath have been discovered, besides antique weapons, sacrificial instruments, and Roman camp kettles of brass. A place called the Castle Hill at Lanark, was the site of a Roman garrison, established there to protect the ford or bridge by which those travellers were wont to cross the Clyde who journeyed along the Military Way which led to the westward. In this respect it was an outpost of the more important establishment at Carstairs. That the Romans had a considerable town at this place is certain, from the great quantities of Roman bricks which have been dug up, remains of buildings which must have stood without its walls, and from the large number of coins discovered. The existence of a Roman bath at this place shows that it was not simply a frequent accompaniment, but a necessary complement of a Roman station.

At Burghead, also, one of their most distant outposts in the north, the ancient *Ptoroton* of the Romans,— which name is still to be recognised in the designation under which it goes among the Highland peasantry in the neighbourhood, *Tory-town*,—vestiges of a Roman Bath have been discovered. Great works, the ruins of which still exist, had evidently been erected there. They were long supposed to be of Danish rather than of Roman construction; and it is quite well known that the Danes, after the time of the Romans, did obtain a footing in the northern part of Scotland. But there can be no doubt that they were built by the Romans, for the purpose of overawing, and, if possible, of eventually subduing, the wild caterans of the north. The circumstances which prove it to have been a Roman settlement are, the discovery of a bath, purely Roman, at Burghead, and also of a deep well, built in the same manner, and with as much regularity as those which have been brought to light within the Roman stations of the south,—a work, indeed, beyond the capabilities either of the aboriginal inhabitants of the country, or of the pirate rovers and sea-kings of Norway and Denmark.

It is by no means so difficult to account for the fact of the custom of the bath falling into abeyance and disuse in these islands, as to account for its decay in ancient Rome. The Romans continued in Great Britain, masters of it, for the first four hundred years of the Christian era. After their retirement from this country in the middle of the fifth century, on the advent of troublous times in the central seat of empire, confusion and insurrection arose in its distant outposts and outlying colonies. In the barbarous ages which ensued, the ancient Roman cities in Great Britain, with their noble buildings, the baths among the rest, were overthrown and destroyed by the different hostile tribes who contended with each other for the possession of the country. In addition, there would always be the natural dislike of a conquered people to retain and perpetuate the peculiar customs of their conquerors. Thus we are furnished with an easy and intelligible explanation of the reason why the Roman Bath did not take root in this country as one of its permanent institutions, and why, after being introduced, it did not continue and flourish. But though such prejudices were natural enough in the case of our barbarous ancestors, surely they are irrational and absurd in the case of us their descendants. We have already derived from the Romans, arts, architecture, laws, and letters : why not now receive from them once more the bath ?

It is from the Turks, however, that we have actually received the bath in modern times. A hundred years ago, the Turkish practice of inoculation for small-pox was introduced into Great Britain, and thereby the first blow was struck at the ravages of that deadly and loathsome disease; and that remedy, followed up and replaced by vaccination, has comparatively exterminated one of the greatest scourges that ever afflicted the human race. And now again we have received from Turkey a most powerful engine, placed at the disposal of medical science—in the Turkish Bath. The revival of the bath is but of recent date. Mr Urquhart, twelve years ago, struck the first spark in "The Pillars of Hercules." Dr Barter, in Ireland, caught it, and fanned it into a blaze; and since then the spark has burst into a flame, which is now rapidly spreading throughout England. In that celebrated work there is a chapter on the Bath, "which," in the author's own words, "if the reader will peruse it with diligence, and apply with care, may prolong his life, fortify his body, diminish his ailments, augment his enjoyments, and improve his temper : then having found something beneficial to himself, he may be prompted to do something to secure the like for his fellow-creatures." A few years after the publication of that book of travels, Mr Urquhart visited Dr Barter's hydropathic establishment at Blarney ; and the doctor, struck with Mr Urquhart's conversation, and delighted with his description of the Turkish Bath, which he afterwards read in " The Pillars of Hercules," wrote to him as follows :—" Your description of the Turkish Bath has electrified me. If you will come down here, and superintend the erection of one, men, money, and materials shall be at your disposal."

Accordingly, the first Turkish Bath was erected about eight years ago, at Blarney, in Ireland. A year later. we find Mr Urquhart building one at Riverside, and introducing it into Lancashire, in England. Mr George Witt, of Knightsbridge, built one a year thereafter in Prince's Terrace, Hyde Park, London, which immediately attracted the attention of the scientific world, and introduced the bath as a great fact to the rank, intellect, and learning of the metropolis. Since then, it has been spreading through the country with marvellous rapidity. Sir John Fife has introduced it, as consulting surgeon, into the Newcastle Infirmary, with marked success. There are already upwards of thirty public baths in London, and, in addition, three large companies have been formed for the establishment of Eastern Baths on correct principles. One has been opened at Brompton, specially for invalids, others are being built in connection with hospitals and lunatic asylums. Several of the leading metropolitan physicians and surgeons have been eagerly testing its merits upon their own persons, and, indeed, have found themselves under the necessity of doing so, in order to be in a position to recommend it to their patients. And although it has not yet been endorsed by the medical profession as a body, the time, it is hoped, is fast approaching, when every poor-house, infirmary, and hospital, as in Turkey, will have a Roman Bath attached to it. Numerous baths are rising up, flourishing and well supported, in many parts of the kingdom, and even in distant dependencies, as in Melbourne and Sydney. The first ever erected in Scotland was the beautiful private Turkish Bath of Andrew Wauchope, Esq., of Niddrie Merschell, near Edinburgh. in 1859. Not only are there baths in Dublin, Cork, Passage, Limerick, and other towns in Ireland; several in Glasgow, one in Aberdeen, and two—my own the first, in Edinburgh; but, besides those already mentioned in London, baths have sprung up in Bradford, Leeds, Manchester, Sheffield, Rochdale, Rotherham, Barnsley, and Staleybridge, and other places in England. In the first of these towns, Bradford, no less than six Turkish Baths have already been erected, and—which must be gratifying to every one who desires the social elevation of the working-classes, the poor man and the artisan can have a good Turkish Bath in that town for the small sum of threepence. What is singular and pleasing in connection with the movement for the revival of the bath in these great manufacturing towns in England, as Dr Wilson observes, is, that, "after the lapse of 1500 years, it should have been so eagerly taken up and adopted by the working-classes. Several baths, founded and maintained by working men, have been established in our great manufacturing towns; and this fact, together with the general popularity of the bath among the artisan class, will doubtless lead the thoughtful man to recognise in the institution attributes of sterling value and prospective public utility." *

It is a fact, however, of which Ireland may justly be proud, that the first Turkish Bath erected in these kingdoms in modern times, and specially designed for

* The Eastern or Turkish Bath, by Erasmus Wilson, F.R.S.

curative purposes, was erected by an Irishman upon Irish ground. The Eastern world had long enjoyed the bath as a social and religious institution—for the Turks gave it a religious character, and, as Disraeli observes, ' both Mohammed and Moses made cleanliness religion." But the shrewd intelligence of Dr Barter first saw a great sanitary principle in it, and straightway set to work to apply the bath to the cure of disease. Mr Urquhart had survived the obloquy poured upon him at first, and now, in the establishment of the first Turkish Bath in modern times on Irish soil, and in the opening of the St Anne's Hill Baths, Blarney, he was privileged at last to see his labours crowned with complete success. But Dr Barter had now for a time to come in for his share of reproach and abuse, and to contend at first with the strongest prejudices and the most violent opposition. To such an extent did this go, that the visitors and patients at the establishment, averaging from 80 to 100, quickly dwindled down to as low as 30. And, on the Doctor's friends remonstrating with him upon his apparent imprudence in risking his well-earned reputation by introducing so great a novelty, he replied in a way truly noble, and worthy of himself. This, he considered, was but a farther development of the principle on which hydropathy is founded—the key-stone of the arch, and, said he, " If every soul leaves my establishment I shall still persevere ; the bath is founded on truth, and must ultimately succeed." The result proved that he was correct. The tide of public opinion, always fickle at best, which had receded so far back, soon turned, and set in strongly in favour of the Oriental Bath ; and to such an extent did this go, that the proprietor was compelled to build additional accommodation, and now there are no less than four sets of baths at St Anne's Hill, and these Temples of Health are steadily increasing all over the three kingdoms. Vast numbers of invalids, particularly in England and Ireland, are availing themselves of this most effectual means of relief and cure. And I have no doubt that they will be equally well patronised in Scotland when the masses of the community are more enlightened, more fully alive to the great value of this boon to suffering humanity, and more completely aroused from that apathetic slumber so proverbially characteristic of the people in reference to the receipt of any new truth—although, in the present instance, the truth submitted is not new, but an old truth, entombed for ages, so far as this country is concerned, and but recently disinterred from the rubbish and debris of nigh two thousand years.

In concluding this sketch of the history of the Roman or Turkish Bath from the most remote to the most recent times, let it never be forgotten that to David Urquhart, Esq., must be given the palm of merit for having introduced it into this country. In urging the duty and the means of personal purification upon the attention of his countrymen, he was assailed for years, as many will recollect, with opprobrium and ridicule, in literary periodicals of the highest character. Ignorant and self-conceited scribes having thus secured the laughter of the country against him and his opinions, the rediscovery of a precious truth seemed in a fair way of being overlooked and perishing. But here we have an illustration of the remark sometimes made, that great truths, although often and long neglected, and even sunk into oblivion and contempt, are never suffered by the Creator to be forgotten, but some one always remains to bear witness to them. All honour then be to Mr Urquhart, who, with praiseworthy zeal, persevered in his noble mission through evil report and through good report, and now at last beholds the triumph of his disinterested efforts. While Dr Barter was the hand, Mr Urquhart was the head. And, indeed, it is pleasant to see Dr Barter fully admitting this, and in frank and manly terms saying— " I willingly and gratefully acknowledge it is to him I stand indebted for directing my attention to this bath, and I shall never cease to be thankful that I was permitted to have the advantage of his knowledge and direction in its construction." And now that the bath has been restored to our island, let us remember that though commonly known as the Turkish Bath, it is not necessarily nor exclusively Turkish at all ; but the ancient Roman Bath, derived originally from the East, coincident with every stage of Eastern civilisation, and at one time co-extensive with the old Roman Empire, having been possessed by the greater portion of Europe, and even by Great Britain herself. In this matter at least we are only going back to the good old times. And let us be thankful that the ancient Roman Bath, after being kept alive for many centuries by the fostering care of the Turks, has at last come back to revisit its ancient haunts, and to offer to us the enjoyments and blessings which our rude forefathers were unable to appreciate.

CHAPTER III.

NATURE OF THE BATH.

There are many different kinds of baths, but the Roman or Turkish Bath is, in its nature, peculiar and singular, and different from all the rest. There are the hot and the cold bath, the warm and the tepid, the vapour and the medicated, the fumigating and the electric, the gaseous bath and the earth bath, the ancient hot-sand bath and the modern mud-bath, the shower and the needle, the douche and the wave-bath; and, counting the natural in addition to the artificial, the fresh-river and salt-water baths. They may all have their respective advantages and virtues when properly used, but in speaking of the Roman Bath, I am under the necessity of premising that the term *bath* is, to a certain degree, misleading. Let the reader dismiss from his mind all preconceived notions of bathing. Let him disabuse himself of all ideas connected with the immersion of the body in the element of water. The Roman or Turkish Bath is not, in the common acceptation of the term, a bath at all. The body is not immersed, at least not necessarily immersed, in water, nor enveloped in watery vapour. It is not, in fact, a bath of water at all. Water may be used, and is used, as being requisite to complete the process. But the principle of the bath is not water, but *Air*—HOT AIR. Neither are you introduced into a steamy atmosphere, but purely and simply into a bath of highly heated air. This is the element of the bath.

Neither are you introduced into a small box or bedstead, as in the vapour-bath, where you can hardly breathe; nor are you and the chair you sit on enveloped in a blanket, as in the lamp-bath, with a burning flame glowing beneath you, and generating carbon all around you. But you are in a large chamber—at least it is to be hoped it is large and well ventilated; and that it is good air you breathe, and not an unwholesome atmosphere from being sunk in some underground cellar. In this chamber there may not be a single drop of water. Indeed, in the Sudatorium, the hottest of all the chambers, there is none. There is water in the first hot-room, no doubt, for quenching thirst, and for washing your feet if you like; and you may get your skin moistened all over if you wish it, or think you require it, before commencing the perspiratory process. When that process, which is the chief business of the bath, is thoroughly completed, then water makes its appearance at last. After entering a succession of different chambers, heated at increasing temperatures as you advance onwards, then, finally, you come to the only place where water is used. There you undergo a washing with water, wave or spray, as you like it best, with soaping and plentiful rubbing down—all which processes I have yet to describe—and by means of which all the excreted matter expelled from the interior of the body, but still adhering to it externally, is effectually removed. Then you retire, last of all, to the cooling-place, where the heated surface of the body regains its normal temperature. The pores of the body, so freely opened in the hot chambers, have been by this time, however, previously closed and gradually contracted in the Lavatorium, by the application first of tepid and then of cold water, so that there is now no danger whatever of catching cold. So that when you come finally into the cooling-place, the Frigidarium, you cool down in an atmosphere which is simply of the ordinary temperature, neither artificially heated nor artificially cooled. But, from first to last, you are never immersed in water of any kind. You are immersed in a bath of hot-air. Heated air is the essence—the secret of the bath.

In the construction of the bath, the most important point to be considered is the manner of producing the required heat. Not only is a high degree of temperature to be produced, but we must also have it under control, so as to be able to regulate it at pleasure, and to render it at all times uniform. It is proper, therefore, that, at the outset, I should say a few words respecting the different modes of heating the bath. The plan of the ancient Romans was by means of the hypocaust, placed directly underneath the hot rooms. This arrangement was doubtless rendered necessary by the vast extent of their bathing establishments. In the construction of some of their smaller Thermæ, however, it would appear that this plan was modified to the extent of placing the hypocaust, or under-ground furnace, a little apart, and introducing the heated air underneath the hot rooms by means of oblong pipes or flues. This plan is attended with the desirable result of having the temperature under complete control. Accordingly, in the reconstruction of the Roman Bath in this country, the heat is produced by a series of hot-air flues underneath the whole of the floor of the hot rooms. These flues or pipes are connected with a furnace on the outside at the one end of them, and a long chimney-stalk at the other, for the purpose of drawing off the flame, smoke, and heated air. These flues are so constructed that they can be properly cleaned and swept; and when they are properly arranged and duly attended to, there is no difficulty in getting the required amount of heat. Thus it will be clearly seen that the principle of the ancient Roman Bath is here faithfully followed out, in the application of the heat underneath the floors of the hot rooms.

Some parties in this country, however, thinking to improve upon the bath of the Romans and Turks, have objected to this arrangement, and have proceeded so far with their so-called improvements as to erect baths of a different construction, and on another principle. Their objection lies against the highly heated state of the floors, in the hottest room especially, which, of course, cannot be entered without wooden clogs. But this is a very trifling disadvantage compared with the great benefits and delights of the bath, and the bather must even consent to keep his pattens on, and walk about in them, for the sake of his *understanding*. Any apparent inconvenience caused by the wearing of the wooden sandals is counterbalanced by the deliciously equal diffusion of the high degree of heat, which the experienced bather so much enjoys and delights in, through every part of the bathing apartment. And this is absolutely necessary to the success, safety, and comfort of the bath. Another counterbalancing advantage is that, by adhering to the ancient Roman and modern Turkish plan of heating the bath, the heat can be regulated with the greatest degree of accuracy.

The plan recently introduced by some modern innovators and improvers is, that of arranging the flues not underneath the floor, but round the sides of the rooms, and immediately under the benches on which the bathers recline. The advocates of this new plan contend that it is a great improvement upon the old, and the chief advantage insisted upon is, that the floors of the heated rooms can be traversed without shoes or clogs. This last point is altogether unworthy of any more serious notice. But, upon the new plan itself, I may observe—and it is the opinion of many well qualified to judge in such matters—that I consider it to present far more serious disadvantages than any attaching to the ancient plan, which, besides the recommendations already stated, possesses also the sanction and experience of ages in its favour. The disadvantages connected with the new plan are, that the heat cannot be equally diffused throughout the whole of the apartment, that it is hotter in one place than in another, and that, while the floor is so cool as to admit of a person walking upon it with bare feet, the concentrated heat under and on the benches, and close to the walls of the apartment, is so great that the bather is in danger of sustaining a severe burn, and this some have already painfully experienced. I visited one of this sort of so-called Turkish Baths in Glasgow, and felt very uncomfortable all the time I was in it from this very cause. In erecting my bath at Sciennes Hill, in this city, I had many interviews with engineers and professional builders, men of tried experience, and much consideration was devoted to the question as to which of the two plans, the ancient or the modern, should be adopted. The matter was fully discussed in all its bearings, and the result was, that the new plan was not sufficiently tested to warrant a departure from the old in its favour.

In proceeding to describe the bath, therefore, I desire to be understood as speaking of the *bona fide* Roman or Turkish Bath, constructed upon sound principles, built upon the ancient plan which for generations has served the purpose so well, and fitted up and ventilated in such a manner as to secure the safety and comfort of invalids, as well as of others who are willing to avail themselves of what, considered only as a luxury, is so profitable and beneficial to the health. For there are many so-called Turkish Baths which are not baths at all, but places got up on the lowest possible scale, and at the least possible expense, merely as a trap for the unwary, and a speculation for obtaining money on false pretences. Many of this class are to be found in London, and in other places, under the superintendence not of medical men, as they ought always to be, but of parties who know nothing of the human frame and its numerous and often complicated diseases, and even less and worse than nothing of their proper treatment. There is as much skill and discrimination required in the proper application of baths of all kinds—the Turkish included—as in the administration of medicinal substances for the cure of disease; and, from ignorant and unprincipled recklessness on this point, much harm has resulted to individuals, and the real and proper Turkish Bath has been, and still is, evil spoken of, and its usefulness, in consequence, greatly impaired.

I now proceed to give a description of the bath, its different processes and apartments. There were no fewer than seven different apartments in the ancient Greek and Roman baths;* but I shall not inflict upon the reader any wearisome description of them, but shall merely observe that all the essential parts of the process are gone through in three, or at the most, four principal apartments. This is the case also among the Turks at the present day, from whom we have in modern times received the bath. They have retained the essentials of the bath, discarding only accessories and superfluities.

The bath with us consists of four apartments. First, the saloon, or divan, furnished with a number of compartments, one of which is assigned to each bather as he arrives. They are elegantly fitted up, and are for the purpose of undressing and dressing in. This we may call the VESTIARIUM of the Romans, or the APODYTERIUM of the Greeks. Here you divest yourself of your apparel, and, having been supplied with a bath dress, which consists merely of a fold of cotton cloth, you adjust it by simply winding it round your loins in kilt fashion. Thus lightly attired, you step into the first hot room, where the temperature ranges from 100 to 120 degrees. This is the TEPIDARIUM. Here you can either

* *Vide* Encyclopædia Britannica—*Baths.*

sit down, recline on a couch, or walk at your pleasure. The warm air is delightful. You may even sleep here if you choose, it is so soothing. " How deliciously," says Dr E. Wilson in his description, " the warm air seems to fold us in its embrace! ' How very nice!' ' How very agreeable!' are the expressions which we hear softly breathed around us. The air is clear—no vapoury mists; it is fresh, for there is a free circulation of air through the room; but how marvellously soothing! All care, all anxiety, all trouble, all memory of the external world and its miserable littleness, is chased from the mind. Our thoughts are absorbed in rapturous contemplation of the delights of the new world—the paradise into which we have just been admitted. The tyrant PAIN, even, loses its miscreant power here; the toothache, where is it gone? the headache, gone too; the spasm no longer bides; the pang of neuralgia, of rheumatism, of gout—all are fled: for this is the region where the suffering find a soothing relief from all their torments. Over the door it is written— PAIN ENTERS NOT HERE!"

The heat in the Tepidarium is by no means oppressive and unpleasant, except it may be on a first visit, and to those only whose skins are not in good condition, but are of a hard, dry, horny-hided nature. And even this condition of the skin is in all cases overcome, and all inconvenience entirely obviated after the first two or three baths. In the generality of cases, the bather begins gently to perspire. His skin becomes warm, soft, and moist. The heat has begun to act on the pores of the skin. In half an hour or so the pores have begun

NATURE OF THE BATH.   51

to act freely, and perspiration bathes the surface of the body. The beads of sweat fall down from your nose and chin, your elbows and finger-points, and all parts of your body. You become "like Niobe, all tears." But unlike all other tears, there is a sense of pleasure and relief connected with the shedding of these copious dewdrops, for as Sanctorius says, "Melancholy is overcome by free perspiration, and cheerfulness, without any evident cause, proceeds from perspiration succeeding well."

You remain in this apartment a longer or shorter period, according as the perspiration comes slowly or freely. Some prefer to take a long and full bath ; others, and the more judicious, prefer to take it moderately and more frequently. And this is in all respects a better and more successful plan ; for excess in this, as in every other indulgence, is to be avoided. A Roman emperor, it is said, killed himself by excessive use of the bath,—the only instance of the kind that we read of. But the most wholesome food, if taken to excess, will have the same effect ; and we have frequently read of people dying of a surfeit. As this, however, is held to be no argument against food, neither is it against the bath. Soon now, and while you are still revelling and luxuriating in the enjoyment of the delightful clime whose mild and warm breath you have been inhaling, the attendant approaches you, and, with practised touch, informs you—ah ! too soon—that you may now enter the next and hottest room.

This is the CALIDARIUM, the hot-room, *par excellence*. It is also called the SUDATORIUM, or the *Sweating-Place*. The bather, in passing from one apartment to another, becomes gradually acclimated to the heat. This room is generally heated to 140 or 145 degrees. Here you remain for fifteen or twenty minutes, and here the perspiration flows very freely. Here also, as well as in the preceding apartment, or Tepidarium, you drink freely of cold or tepid water, with which you are supplied by an attendant, according to your wish and your desire for it. And it is well to do so, as being useful in replacing the aqueous part of the perspiration. Thirst is not so much felt by the inexperienced as it is by the practised bather who learns to drink plentifully and with pleasure, until the element which he imbibes passes out of the pores of the skin as clear as when taken in at the mouth. Not till then can you be said to be thoroughly cleansed.

This room, the Sudatorium, is the climax of enjoyment. It is not indeed so decidedly so at the first, or perhaps even at the second or third bath. If any inconvenience is felt at first, let the bather, instead of reclining on the heated couch, sit up, or, better still, walk about in the deliciously pure and warm air, and his sensations will be most rapturously pleasant. After a few visits, the constant—say the weekly—bather longs for this apartment. After he has been in the Tepidarium half an hour or forty minutes, he begins ardently to desire the greater warmth of the Sudatorium. That which on a first visit seemed to catch his breathing shortly, and appeared unbearable for any length of time—caused by his skin not acting so freely and responding so readily to the heated air as it has since learned to do—now seems to encircle and enwrap his whole person in waving masses and volumes of rich balmy breath.

It is here that the full benefit of the bath as an eliminator is experienced. The pores of the skin are now freely opened and a copious perspiration is induced. And not only are the pores now opened, but they are thoroughly cleansed and relieved, and thus all the foul, effete, superfluous, and noxious matters that were lodged in them, and in the tissues generally, are entirely ejected.

Perhaps, however, it may be supposed that so great a perspiration must disturb the economy of the system. But the drain is only for a limited period. It is not continued so long as to be in any way exhausting. And besides, as Dr Cumming says in the Dublin Hospital Gazette, " there is always at hand the means of supplying the waste, and preserving intact the specific gravity of the blood, by draughts of cold water, which is taken up directly by the veins of the stomach, thrown into the circulation, and thrown out from it again in the shape of perspiration. all within the space of a few minutes." Moreover, as is well known, healthy perspiration is never weakening. A man who walks a long distance, or digs in a ditch all day, necessarily perspires a good deal, but the loss of tissue is quickly made up afterwards by abundance of nourishing food. Thus the liquids of his body that were drained off are supplied, its solids maintained, and the standard of his physical strength is not only kept up but increased by combined healthy exercise, pure air, and good nourishment. For the appetite being rendered vigorous, the food taken is rapidly assimilated. The case is the same with the perspiration artificially produced by the Turkish Bath, and their general result in like manner is, that the body is really and ultimately strengthened. It is different, however, with the perspirations produced by disease. Copious night sweats are weakening and disturbing in their effects, as the debility and disease from which the patient suffers prevent him from assimilating the nutriment which he swallows. But the testimony of all who have made a fair trial of the Roman or Turkish Bath goes to establish the fact, that, so far from producing any derangement or disturbance of the internal economy of the body, it is the blissful cause of exciting an admirable and harmonious working of all its functions.

Before leaving the Sudatorium, and while the perspiration is flowing freely from every pore, the shampooing process is performed. This operation is not attended with any unpleasant sensations, but is conducted in a way that can cause no uneasiness even to the most sensitive person. If desired, it may be omitted altogether. But in the opinion of most persons who have considered the subject, either it, or—in the absence of it—the glove constitutes a very essential part of the process. The surface of the body having become warm and moist, and the layers of scarf skin on the surface having become softened and swollen, and ready to be peeled and rubbed of, it is necessary that it should be removed by friction. By this means only can the filth of the body, brought to the surface by the heat and perspiration, be effectually relegated, and the circulation equalized over the whole body. It consists of a gentle kneading, rubbing, and pinching of the limbs, back, and chest, together with a stretching of the joints. The accumulated deposits of insensible perspiration become, under this process, freely disengaged from the pores of the skin, and, combining together, fall from the body in rolls like shreds or pellicles of paste, which, to the uninitiated, might seem like the peeling off of the outer cuticle itself. It is, in reality, however, only the getting rid of the impurities that have been long collecting on the surface of the body. This phenomenon is witnessed particularly in the case of those who take the bath for the first time. The quantity of dirt forced out from the pores, and of dead skin and effete matter removed from the body in this, the manipulating part of the process, is truly astonishing; and it is really wonderful how any tolerable degree of health can be enjoyed with such an artificial covering on the surface of the skin.

In describing this part of the process, which appears very formidable as practised in Turkey, Mr Urquhart talks of the *Tellak*, or bath attendant, throwing himself on the bather's body, griping, hauling, pushing, contorting, jerking, doubling him up, and other gymnastics, all which can only be pleasantly exaggerated figures of speech. But he goes on to give a description of the actual results of the rubbing and shampooing :—" The *Tellak* puts on the glove of camel's hair. He stands over you; you bend down to him, and he commences from the nape of the neck in long sweeps down the back till he has started the skin; he coaxes it into rolls, keeping them in and up till within his hand they gather volume and length; he then successively strikes and brushes them away, and they fall right and left as if spilt from a dish of macaroni. The dead matter which will accumulate in a week, forms when dry a ball of the size of the fist. I once collected it and had it dried—it is like a ball of chalk."

When I first read this, I confess I was rather sceptical as to the trustworthiness of the statement, but a very short experience of the bath on my own person, as well as on many others, proved to demonstration the truth of Mr Urquhart's assertion. I have repeatedly seen parties, even in the Tepidarium, exude such a quantity of effete debris from their bodies, that the white sheet on which they reclined frequently had the appearance as if one had taken a handful of macaroni, and strewed it along on each side of where they lay. Shampooing, then, is a beneficial, healthful, and invigorating process, causing the blood to flow briskly through the capillaries and smaller veins, and bringing all the muscles and sinews of the body into active play, and without the fatigue of severe exercise. It also brings the pores into a state of healthy activity; and few who have once undergone the operation, would consider that they had received the full benefit of the bath were it omitted.

This having been concluded, the bather is now conducted to the LAVATORIUM. Here there is water. Here the bather, after being well soaped with a glove of another sort in the attendant's hand, is thoroughly washed by an abundance of water, which is made to fly out from the wall overhead in all directions on his body, in the form of gentle spray, douche, or wave, according to taste, and of any desired temperature, generally beginning with warm, then proceeding to tepid, and ending with the cold spray. This is the culminating point of the bath. The sensations experienced here are exquisitely delightful; and it is impossible for any one who has not experienced it, to conceive the pleasure afforded by this part of the process. It is indeed a new sensation that is experienced here. How enjoyable to the bather is the warm cascade which bursts over him, as soon as the soaping and friction are ended! How difficult it is for him to bring his mind to the belief that he has had enough. But if the sensation produced by the warm shower is agreeable, no less pleasant is the tepid, but infinitely more so is that by which it is immediately succeeded, namely, the spray of coldest water which, like the warm spray, is made to play all over the body. Its coldness is now most grateful, its enlivening freshness most delightful; and there is not the least danger of striking a chill, as the inexperienced might imagine, the body having been so thoroughly warmed up by the preceding operations. The contentment, pleasure, and delight experienced in the *Lavatorium* is undoubtedly greater than in any other part of the whole process, and the only regret felt is, that an enjoyment so delicious should come to an end. I have heard a bather, a learned Professor in the University, on emerging from the Lavatorium, exclaim—" Glorious!" And one feels as if he could put in for the reward offered in the proclamation of an ancient king for the discovery of a new pleasure, and cry out—as the philosopher did of old, when he hit upon the discovery of a long sought for truth—" Eureka! I have found it!"

In the above brief description of the different processes of the bath, so far as they have yet gone,—and they are now nearly ended, it will have been observed that until the bather has arrived at the Lavatorium, soap has not touched his skin. The moistening of the epidermis in the Tepidarium, and the bursting forth and free flow of the perspiration in the Sudatorium, are left to the operation of nature, under the influence of a high degree of heat. The shampooing, too, is carried on without the aid of soap. Strange to say, the use of soap would have interfered with, and materially spoiled, the success of that operation, as any one may find by experience. The explanation probably is, that the alkali in the soap, by combining with the oily material in the dead scarf, which is to be peeled and rubbed off by friction from the body, deprives it of the consistency necessary to its easy removal. But, after you have been thoroughly sweated, artistically rubbed down and kneaded in the Sudatorium, and the dead epiderm has been completely softened and detached—in a word, after you have been done to a turn—then you are ready for the Lavatorium, and now new wonders are to be revealed. The old scarf has been shed, but the exuviæ are still adhering to all parts of your body, and it is now that soap comes into play.

After a good lathering with soap, and a thorough ablution with water, the softened layers and particles of dead skin still adhering slide off to the ground, and leave behind them, newly developed, the pure, bright, and satiny surface beneath. You touch your new skin with a feeling of wonder and pleasure. All the extraneous atoms and foul scales which clogged and impeded the healthy working of the innumerable pores of the body have been washed away by rushing jets and streaming showers of deliciously warm water. The spray alternates, according to your desire, from hot to cold, and from cold to hot. But be it always remembered, that the constant stream ever flows over the person from a fresh source, and is then carried away by conduits from the floor; that the water, in a word, is always pure, and is never allowed to accumulate in any vessel in which the body, or any part of the body, is immersed. Indeed there is no such immersion; and water once used is never used again, but all is fresh and pure. And this constant flow, and rapid alternation from hot to cold spray, is so delightful to the invigorated and rehabilitated bather that, instead of shuddering and shivering when the cold jet is playing over him, he feels such a degree of comfort that it is difficult for him to say at times whether it is the hot or the cold shower that is gently bedewing him. Yea, his internal feeling of comfort is now so great, and his delight in the tonic and bracing cold shower so extreme, that I have even heard a bather shout out, amid the noise of the rushing waters, "Colder! colder!"—while the attendant was all the while assuring him that he could give it to him no colder, and that it was of the natural temperature, and this was in the depth of winter. This final and concluding cold affusion cools the heated frame, and produces contraction of the mouths of the seven millions of pores which open on the surface of the skin. These pores have acted freely, they have performed the duty that was required of them. They are now in a state of healthy, natural, and vigorous action.

Last of all, he is conducted into the FRIGIDARIUM, or the *Cooling-Room*. A dry warm sheet is wrapped around him, enveloped in which he retires to a couch, and reposes until the heat of the body is gradually reduced, and his frame resumes its normal and natural temperature. These large and comfortable sheets are manufactured for the purpose, and are admirably adapted to absorb moisture. Any slight feeling of languor consequent on the manipulations to which the bather has been subjected, which, however, is very rarely felt, is now followed by a most agreeable feeling as of renovated existence. You either lie down on the soft couch, or sit up, according to your taste, with your sheet thrown loosely and lightly about you. Oxygen is now freely absorbed through the skin, the body being fully exposed to the action of the air, while the pores are in the best condition to receive it. The air in the Frigidarium is of the natural temperature.

This is the true luxurious period of the bath. A sense of perfect ease and unlimited satisfaction is felt. A cup of coffee is sipped *a-la-Ture*, if desired. In this apartment you remain fifteen or twenty minutes, as long, in fact, as you feel it agreeable, till you are cooled, but not cold, and then the returning desire for exertion is the best proof that the bath is completed. It is well to bear in mind that in this, the final part of the process, you are still bathing—bathing now in the cool *air*, and therefore it is well not to wrap up too much here, but rather to expose the body as much as possible to the fresh air. And there is no fear whatever of catching cold: the change of temperature is in fact grateful to the feelings as well as salutary to the body.

The whole process occupies about an hour and a half. Everything is quieting to the feelings, luxurious and tranquillizing. Everything you touch is clean, the water is fresh, the atmosphere is pure, and the whole process is purifying and strengthening. The bather now seeks his own compartment in the saloon, performs his toilet, and quits the bath light, cheerful, buoyant and happy. Staid old gentlemen, grave and reverend seigniors who have not performed a caper perhaps since the days of their hot youth, the dyspeptic, the rheumatic, and valetudinarians of all sorts, now feel as if they could skip and jump. The step is light and elastic, the walk firm, the nerves at ease, the mind cheerful. The bather feels like a giant refreshed with wine, and as if a great weight had been removed from his system,—and the world as a whole seems rather better than it was before. The calls of appetite begin to make themselves felt; nutritive fuel is loudly called for to feed the constitutional fire; and the grateful bather, with relieved, refreshed, and restored system, thanks Mr David Urquhart for the reintroduction into this country of the Roman Bath, and begins to think with the poet,—

"Oh! if there be an Elysium on earth, it is this! it is this!"

I cannot refrain from introducing in this place Mr Urquhart's animated description of the effects and sensations produced by the bath. He says, "The body has come forth shining like alabaster, fragrant as the cistus, sleek as satin, and soft as velvet. The touch of your own skin is electric. Buffon has a wonderful description of Adam's surprise and delight at his first touch of himself. It is the description of the human sense when the body is brought back to its purity. The body thus renewed, the spirit wanders abroad, and, reviewing its tenement, rejoices to find it clean and tranquil. There is an intoxication or dream that lifts you out of the flesh, and yet a sense of life and consciousness that spreads through every member. Each breastful of air seems to pass, not to the heart, but to the brain, and to quench not the pulsations of the one but the fancies of the other. That exaltation which requires the slumber of the senses,—that vividness of sense which drowns the visions of the spirit,—are simultaneously engaged in calm and unspeakable luxury: you condense the pleasures of many scenes, and enjoy in an hour the existence of years. But this too will pass. The visions fade, the speed of the blood thickens, the breath of the pores is checked, the crispness of the skin returns, the fountains of strength are opened; you seek again the world and its toils; and those who experience these effects and vicissitudes for the first time exclaim, 'I feel as if I could leap over the moon.' Paying your pence according to the tariff of your deserts, you walk forth a king."

In Chapman and Hall's "Library of Travel" there is a similar, though briefer description. After stating that the Orientals are passionately fond of the bath, and that every European who has tried it speaks with much satisfaction of the result, the writer goes on to say, "When all is done, a soft and luxurious feeling spreads itself over your body; every limb is light and free as air; the marble-like smoothness of the skin is delightful; and after all this pommelling, scrubbing, racking, parboiling, and perspiring, you feel more enjoyment than ever you felt before." Dr Erasmus Wilson's description, as coming from one of the medical profession, is also worthy of attention. After being urged by a friend to try the bath, he says, "When I stepped into the Calidarium for the first time; when I experienced the soothing warmth of the atmosphere; when, afterwards, I perceived the gradual thaw of the rigid frame, the softening of the flesh, the moistening of the skin, the rest of the stretched cords of the nervous system, the abatement of aches and pains, the removal of fatigue, and the calm flow of imagination and thought,—I understood the meaning of my friend's zeal, and I discovered that there was one bath that deserved to be set apart from the rest,—the bath that cleanses the inward as well as the outward man, that is applicable to every age, that is adapted to make health healthier, and alleviate disease, whatever its stage or severity."

An interesting question now arises for our consideration, namely, How far does the present Turkish Bath agree with or differ from the Roman Bath? To which may be added another important inquiry, namely, In how far is the bath, as now introduced into this country, founded upon the ancient Roman model? Taking the last of these two questions first, it may be answered, that the bath, as it has now been described in these pages, is, so far as can be gathered from the writings of ancient Roman authors, substantially the bath of the Romans. They had all the different apartments and processes already described. Galen says, that "on entering, they remain in the hot air, and are afterwards washed first in hot water, and finally in cold water, wiping off the sweat; and those who do not go from the *Sudatory* into cold water, are in danger of bursting out on returning to the dressing-room into a second sweat, which leaves them chilly." Even to minute points the resemblance, nay, the identity of the two processes can be traced. The shampooer, for instance, was called the *tractator*. They had indeed, in the Lavatorium, a *labrum*, or large marble trough, on the edge of which they sat, and were then rinsed with warm water poured over them by means of a cup or basin, the *pelvis*. But while the spray or douche is in all respects a more effectual mode of washing, there can be no doubt that the Romans also had streams of pure running water from above. This is clearly seen from the plan and drawing of the Roman Thermæ on the walls of the baths of Titus, and from innumerable indications in the ruins of their baths of the use of pipes, plugs, and cocks. They had also the towels or rubbing cloths, called *lintea*.

One apartment in the Roman Bath has disappeared in the modern process, and also in the present Turkish Bath. I refer to the UNCTUARIUM, or *Anointing-Room*, into which the bather passed from the Frigidarium, and where he was smeared with fragrant oils before resuming his dress in the Vestiarium. The attendants here were called *Unctores*, their vial or cruet of oil *guttus*, generally made of horn, *corneus*, and hence a large horn is in Juvenal called *Rhinoceros*. It was this custom which rendered the use of the *strigilis*, scraper, or curry-comb necessary to rub off the greasy sweat and filth from their bodies. The strigil was usually of horn, sometimes of brass or iron, and even of silver or gold. As the Romans used neither linen nor stockings, and wore loose flowing garments, we see a reason for them fortifying their bodies with oils and ointments. To the Turks the use of these things would naturally be abhorrent, for they were taught by their religion to regard animal oils as a defilement, not only after the purification of the Bath, but at all times. And as regards ourselves, we seem generally agreed in discarding the use of bodily anointing, from the apparent uncleanliness of the custom; and also, because the abundance of tight, warm, close-fitting garments which we wear, appears to render it unnecessary. Still, it is the opinion of some, that a thimbleful of perfumed olive oil rubbed over the body after the bath, would be both comfortable and beneficial.

As regards the modern baths of Turkey, and their similarity to the ancient baths of Rome, a few observations may not be out of place. Though identical in principle, and substantially the same in the great ends aimed at and accomplished in both, yet there are a few points of difference between them. In both there is, first, the seasoning of the body, next the manipulation of the muscles, then the peeling of the scarf-skin, and finally the washing of the body, and the period of repose previous to retiring from the bath. But

E

among the Turks this final process of cooling the body is much protracted beyond what it was among the ancient Romans, and also beyond what it is among us who have reintroduced the Roman Bath in its integrity into this country. The Romans concluded the bath as we now do with the cold douche, which closes the pores and arrests the perspiration; the Turks omit the cold affusion except to the feet. They retire from the bath heated and still perspiring, and hence they require a change of linen while slowly and gradually cooling, and also the aid of an attendant to cool the body with a fan. We can thus comprehend how it is that the bath in Turkey may be prolonged to two or three, or even more hours. There can be no doubt, however, that the more we revert to the method of the Romans the better is the bath; for, by concluding it with the cold douche or the cold plunge (they used both), the profuse perspiration is immediately arrested by the closing of the pores, no change of linen is required, no fanning is needed, the cooling of the body is accomplished equally well, and in a shorter space of time, and all danger of catching cold is avoided.

Mr Urquhart has noted some other differences between the practice of the Romans and the Turks. "The Romans," he says, "used the bath to excess, taking it daily; the Mussulmans restricted its use to once a-week. The Romans entered the bath naked; the Mussulmans have introduced a bathing-costume. The Romans allowed the two sexes to enter promiscuously; the Mussulmans have wholly separated them. They have preserved the good, and purified it from excess." But serious exception may be taken to this picture which Mr Urquhart has drawn of the practice of the Romans. Seneca, while admitting that the custom of the bath had in later times been carried to an immoderate excess, distinctly says, that their forefathers, in Scipio's days, entered the sweating-places only on market-days. And there was certainly a distinction of the sexes observed, for both Varro and Vitruvius mention the *balnea virilia et muliebria*.

But the principal change which it is acknowledged the Turks have made is the substitution of moist vapour for dry heated air—*siccus calor*, a dry heat, according to CELSUS—as used by the Romans, and generally followed in this country. This circumstance has caused a good deal of discussion among medical men and others, some contending for the moist and others for the dry hot air. The vapour in the baths in Turkey is produced by the number of bathers all throwing hot water upon themselves and upon the heated floor so that it rises up in the form of visible steam, so that in the course of the day visitors from Europe, on going into the baths in Egypt and Turkey, feel stifled, and can scarcely breathe at first, on account of the heat and vapour. Now the presence of vapour betokens a low temperature in the bath, because watery vapour is scalding at 120 degrees of heat, and hence the Turks who permit the air to become vaporised are compelled to restrict themselves to comparatively moderate temperatures, while the Romans, by excluding watery vapour in the hot rooms, were able to introduce into them a high degree of heat, and thus the object of the bath was more effectually carried out. Some, however, consider a certain amount of moisture in the air an improvement to the bath, especially to those who enter it for the first time, as thereby the skin is softened and moistened, and more easily trained to discharge its duty and perspire freely. But this object can be attained as effectually by wetting the body with warm water, and thus a sufficient amount of moisture for the purpose will be evolved under the evaporation consequent thereupon. The atmosphere should never be rendered artificially and perfectly dry, otherwise it would cause much distress to some persons who are slow to perspire. But it is never in any case thus absolutely and perfectly dry, because a certain amount of vaporization must necessarily take place from the recesses in which water is contained for the bathers' use, externally and internally. To this extent also the ancient Roman Bath must have had a certain degree of moisture suspended in its otherwise pure, clear atmosphere, and from the same source, namely, natural vaporization from water, but without any visible haze or steamy vapour, as in the modern Turkish Bath.

Here also the ancient bath of the Romans should be our model. For that bath is unquestionably the best which contains only just a sufficient amount of moisture in the air to render it agreeable to the lungs. For if it were unnaturally dried air, it would be acrid and hurtful to the organs of breathing. And if, on the other hand, it be loaded with moisture and rendered oppressively steamy, then it will produce, if very hot, an unbearable feeling of stifling and exhaustion. Every one has experienced the fact, that a degree of heaviness and moisture in the air not only aggravates the benumbing influence of cold in winter, but also increases the oppressiveness of heat in summer; and every pedestrian knows that he can travel much farther, and with greater ease to himself, in clear, dry, hot weather, than on a close, humid, sultry day. All who have tried the ordinary vapour bath, too, must have felt how intolerable is the feeling of faintness and suffocation experienced whilst the body is enclosed in the vapour box, or vapour chamber, even though the head of the patient is exposed to the influence of the external air ; and on this account also, no one is able to remain longer than a quarter of an hour in it. The reason is that, where the air is saturated with moisture, evaporation from the skin cannot take place. For watery vapour and steam, being denser media than air, interfere with the free evaporation from the body, and prevent the free discharge of effete matters through the skin, by keeping the pores comparatively closed. And there can be no doubt that the steam which condenses on the body in the pure vapour-bath is often erroneously mistaken for perspiration. Hence the inferiority of the vapour-bath to the hot-air or Roman Bath. In the lamp-bath, too, the patient soon begins to feel oppressed and faint as if short of breath, because the lamp consumes the oxygen around the body, and the skin can no longer breathe. And not only is perspiration impeded, and respiration by the skin arrested, as in the vapour-bath, but, worse still, the lamp-bath soon becomes a carbonic acid gas bath,—a poison-bath, in fact. But from all these evils, the Roman Bath of pure, natural, sweet, and wholesome hot air, is entirely free. Man was made to breathe air, not vapour. And the Roman Bath shows its infinite superiority to all others, not only by its extraordinary salutary and beneficial effects, as I shall yet show, but also by the fact that it may be enjoyed, even at high temperatures—which is impossible in vapour baths—for more than an hour at a time, not merely without inconvenience but with positive pleasure.

CHAPTER IV.

THE BATH IN DIFFERENT COUNTRIES.

I DESIGN in this chapter to show that the bath is not the singular thing in principle and in practice which many, through ignorance, suppose it to be; and that— that for whose introduction and adoption among ourselves I so earnestly plead, is already and substantially known and valued very widely throughout the world. If we inquire, then, where this bath is to be found at the present day, whether as the simple sudorific bath mentioned by Herodotus among the Greeks, or as the bath of hot air and hot water among the Romans, or as the bath of hot air, hot vapour, hot water and soap, all combined together, as among the Turks, we shall be astonished at the number of different countries of the globe in which it has taken root, and in which its practice, more or less, prevails among the people. Why it should so extensively have fallen into disuse among most of the nations composing the old Roman Empire, while it should still prevail among many nations that were never subject to their sway, may be an interesting inquiry. But I shall not dwell upon that, but proceed to notice the remarkable fact, that, in very many and widely remote regions of the globe, the results aimed at, and the benefits sought to be obtained by this bath, are arrived at by substantially the same means. The survey will furnish us with facts showing its general usefulness and its great benefits as a prophylactic and curative agent, and prepare the way for a consideration of its physiological action, uses, and therapeutic results.

It seems natural, in taking this wide survey, that we should begin nearest home—with Ireland. This country, which contains so many antiquarian remains and mysterious ruins, whose history and purpose are unknown, has possessed what the Irish call "sweating-houses" from time immemorial. They are of the rudest possible construction, and the Irish name for them is TIG ALLUI. A lady happened to mention to Mr Urquhart a "practice of sweating" among the peasantry of Rathlin, in Ireland; and the following is the particular account which she gives in answer to his inquiries:—"With respect to the sweating-houses, as they are called, I remember about forty years ago seeing one in the island of Rathlin, and shall try to give you a description of it. It was built of basalt stones, very much in the shape of a bee-hive, with a row of stones inside for the person to sit on when undergoing the operation. There was a hole at the top, and one near the ground where the person crept in, and seated him or herself; the stones having been heated in the same way as an oven for baking bread is; the hole on the top being covered with a sod while being heated, but, I suppose, removed to admit the person to breathe. Before entering, the patient was stripped quite naked, and, on coming out, dressed in the open air. The process was reckoned a sovereign cure for rheumatism, and all sorts of aches and pains."

Similar primitive hot-air baths, used by the native Irish, have been discovered in other parts of the island,—in Fermanagh, Leitrim, Tipperary, and Tyrone. Two varieties of the *Tig Allui* have been found to exist in Ireland, some of which are in ruins, and others are in actual use among the peasantry at this very day. One kind was tolerably large, and capable of containing a good many persons; and of this the Rev. Mr Gage of Rathlin Island gives the following description:—
" When the people are attacked with rheumatic pains, they have recourse to a remedy of long standing, in the efficacy of which they have great confidence. In several parts of the island small buildings, called *sweat-houses*, are erected with stones and turf, neatly put together, in the shape of a bee-hive. There is a small hole in the roof, and an aperture below, just large enough to admit one person on hands and knees. When required for use, a large fire is lighted in the middle of the floor, and allowed to burn out, by which time the house has been thoroughly heated; the ashes are then swept away, and the patient goes in, having first taken off his clothes, with the exception of his under garment, which he hands to a friend outside. The hole in the roof is then covered with a flat stone, and the entrance is also closed up with sods, to prevent the admission of air. The patient remains within until he begins to perspire copiously, when (if young and strong) he plunges into the sea, but the aged or weak retire to bed for a few hours. This primitive bath has been successful in removing

pains of long standing; and people on the mainland have come for the express purpose of trying its efficacy. It is not, however, applied exclusively to the cure of disease, for the young women not unfrequently resort to it, after burning kelp, to clear their complexions, especially if it should happen to be near the time of the Ballycastle Fair. As the heating of the sweat-house is considered rather expensive, from the quantity of fuel required, a number of persons generally go in together. The period of remaining in is at the discretion of the patients; but at the end of half an hour the house is generally cleared, and the fresh air once more admitted, until its services are once more required."

The description of the second kind of native Irish sweating-house, intended only for a single occupant, is given in a letter by Dr Tucker of Sligo, in the " Fermanagh Mail," as follows:—" It is built of stone and mortar, and brought to a round top. It is sufficiently large for one person to sit on a chair inside, the door being merely large enough to admit a person on his hands and knees. When any of the old people of the neighbourhood, men or women, are seized with pains, they at once have recourse to the sweat-house, which is brought to the proper temperature by placing therein a large turf fire, after the manner of an oven, which is left until it is burned quite down, the door being a flat stone, and air-tight, the roof, or outside of the house, being covered with clay, to the depth of about a foot, to prevent the least escape of heat. When the remains of the fire are taken out, the floor is strewn with green rushes, and the person to be cured is escorted to the bath by a second person, carrying a pair of blankets. The invalid having crept in, plants himself or herself in a chair, and there remains until the perspiration rolls off in large drops. When sufficiently operated on, he or she, as the case may be, is anxious to get out, and the person in waiting swaddles him up in the blankets, and off home, and then to bed. I have heard old people saying, that they would not have been alive twenty years ago, only for the sweating-house." Dr Haughton also says that a few years ago, a rude kind of bath was used in a similar way, near the slate quarries of Kilkenny, heated in the same manner as a baker's oven, and that similar structures, built near running streams, exist in other parts of the country. He also very well observes, that "the bath used by the Spartans, although heated in a different way, very much resembled that which is now used in Rathlin Island; and I am sure nobody can say that it had the effect of debilitating *their* constitutions, or introducing amongst them a hurtful degree of luxury. On the contrary, there is every reason to believe that their great physical prowess and undaunted energy was, in some degree, owing to their attention to the state of the skin."

Among the Mandingoes in the interior of Africa, it is the custom, as we read in Mungo Park's Travels, that, on the first attack of fever, when the patient complains of cold, he is placed in a similar sort of bath. "This is done by spreading branches upon hot wood embers, and laying the patient upon them wrapped up in a large cotton cloth. Water is then sprinkled upon the branches, which, descending to the hot embers, soon covers the patient with a cloud of vapour, in which he is allowed to remain till the embers are almost extinguished. This practice commonly produces a profuse perspiration, and wonderfully relieves the sufferer."

Again, in Gent's "History of Virginia," we read of a similar kind of bath among the North American Indians, namely—a combination of hot air and hot vapour. "The medicine-man, or doctor, takes three or four large stones, which, after being heated red hot, he places in the middle of the stove. This being done, the Indians creep in, six or eight at a time, or as many as the place will hold, and then close up the mouth of the stove, which is usually made like an oven in some bank near the water side. The doctor then pours cold water on the stones, and after the men have sweat as long as they can well endure it, they sally out, and (though it be in the depth of winter) forthwith plunge themselves over head and ears in cold water, which instantly closes up the pores and prevents them from taking cold."

M'Cormack, in his history of the Red Indians of Newfoundland, says that he discovered the remains of a similar bath among them. "The method they took," he says, "was to heat large stones in the open air, by igniting a quantity of wood around them. After this, the ashes were removed, and an hemispherical framework, closely covered with skins, was fixed over the stones. The patient having provided himself with a small bucket containing water, and a bark-dish to dip it out, crept in under the enclosure, and then by pouring the water upon the heated stones, was enabled to raise the steam at pleasure." And Mactaggart, in his "Three Years in Canada," says the same thing "is in great request in the towns of Canada, and found of much service during winter, when the cold seals the pores and checks perspiration. They build the bath of rude stones, by the banks of a lake or river, and kindle a fire and keep it up until the stones be hot; they then sprinkle some water, and bring forth the patient, having stretched him or her in the rude bath; water is poured against the hot stones, which flies hissing on the body; when this is done, it is wrapped up in buffalo skins, and a profuse sweat is hereby obtained."

Ellis, in his "Journal of an Embassy to China," says that at Nanking there is a public vapour-bath, called "The Bath of Fragrant Water," where the Chinese are stewed clean for ten tseen, or three farthings. The bath is a small room of one hundred feet area, divided into four compartments, and paved with coarse marble. The heat is considerable, and the number admitted into the bath has no limit but the capacity of the area. The traveller sneers at the whole thing, says that the stench was excessive, as it was very likely to be from such an indiscriminate crowd, and concludes by pronouncing it "the most disgusting cleansing apparatus he had ever seen, and worthy of this nasty nation—the Chinese." But, as Dr Wilson justly observes, "what would Mr Ellis say of a country in which there existed no 'cleansing apparatus' whatever?—for example, his native country." It is worthy of remark that the Chinese mode of heating their baths, and even their sleeping apartments, as we learn from other travellers, is by means of an underground oven—a hypocaust, in fact, like the Romans.

These remarkable facts, which I have thus produced and grouped together, drawn as they are from widely distant portions of the world, must have arrested the attention and impressed the mind of the most cursory reader, and must go far to produce in every candid person the conviction that the bath of which I am speaking is at once an instinct and a necessity of man's nature. No doubt there is a difference between the bath in these countries and the Turkish and Roman Baths; but after all it is merely a question of more or less vapour in them. Some of those which have now been quoted, are styled by the travellers who visited them vapour-baths; but it is evident, at a glance, that they are very different from the things which go by that name, such as vapour-boxes, vapour-beds, and vapour-chambers, in this country. The disadvantages of such baths of pure vapour or steam have already been pointed out; and it may safely be said that now, since the introduction of the Roman Bath into this country, their day is over. But in most, if not all, of those which have now been brought forward, as practised by the people of different countries, it is clear that there is present, besides the vapour, which they knew not how to exclude, plenty of air, and hot air too; so that in them, as in the present Turkish Baths, it is apparent that there is the same combination of hot air and vapour. This fact, however, does not detract in the least from the justness of the principle of the bath, nor in any great degree from its value and its benefits to the people who practise it. The presence of the vapour will doubtless, at high degrees of heat, be attended with inconveniences to the bathers themselves, and the same result will follow from the rudeness of their contrivances in general. But the grand result aimed at in all, is accomplished in all, and—*finis coronat opus.* The only difference between them and a scientifically constructed Roman Bath is, that in it the result is accomplished more successfully, more comfortably, and more safely. And the conclusion to which we come is, that when we find the Red Indian, the Mandingo, the Irish peasant, and the Chinese coolie, all well acquainted with the virtues of heated air, and frequently having recourse to it with the happiest results, its antiquity as a remedial agent, however discovered, or from whence derived, and its coincidence with every state of society, the most polished and the most barbarous, clearly prove it to be an instinct of nature.

This conclusion, together with its salutary effects in a physiological and medical point of view, will become only more clearly apparent from a farther collation of facts. In prosecuting our survey, then, I now proceed to speak of the bath in those countries into which its introduction can be more easily accounted for. Kohl, in his work on Russia, thus describes the Russian Bath:—" On Sunday evening an unusual movement may be seen among the lower classes in St Petersburg ; whole companies of poor soldiers who have got a temporary furlough, troops of mechanics and labourers, whole families, men, women, and children, are eagerly traversing the streets with towels under their arms, and birch twigs in their hands. From the zeal and haste manifested in their movements they would seem to be engaged on important business, as, in fact, they are—the most important and agreeable of the week. They are going to the Baths, to forget in the enjoyment of its vapours the sufferings of the past week, to make supple the limbs stiffened with past toil, and to invigorate them for that which is to come. The Russians are such lovers of vapour baths that St Petersburg contains an immense number of these establishments. Before the door, the words 'Entrance to the Baths,' in large letters, invite the eye. Within the door-way, so narrow that only one at a time can work his way, sits the money-taker, who exchanges a ticket for the bath for a few copecks. Men, women, boys, and girls, all hurry to secure their tickets, as if proceeding to some favourite show. The passage is divided into two behind the check-taker's post, one for the male, and one for the female guests. We first enter an open space in which a number of men are sitting in a state of nudity on benches, all dripping with water and perspiration, and as red as lobsters, breathing deep, sighing, puffing, and gasping; and others busily employed in drying themselves and dressing. These have already bathed, and now, in a glow of pleasurable excitement, are puffing and blowing like Tritons in the sea. Even in the winter I have seen these people all melting from the hot baths, drying and dressing in the open air, or at most in a sort of booth forming an outhouse to the bath."

Dr Robert Lyall's description of the Russian Baths is

still more animated and suggestive. After mentioning that the heat is generated by means of a large stove filled with stones, which are heated by means of burning wood, over which water is thrown, he says. "The steam thence raised fills the apartment, which is surrounded with a wooden platform of ascending stairs or steps, and, according to the degree of heat desired, the person ascends the higher. The first sensations on entering this apartment are very singular, chiefly from the difficulty of breathing such a hot and moist atmosphere. On the platforms, which are raised in the form of an amphitheatre, lie an immense number of persons apparently inflicting torture on themselves. If not dead, they actually seem struggling with death, for the air they are breathing can only serve to stifle. Other persons, their tormentors, are employed in scourging them with birchen rods, steeped in cold water, as if to increase the smart. Others, standing by the glowing stones, and steaming at every pore, have ice-cold water poured over them by pailfuls. When the first disagreeable effect of the heat is overcome, and the transpiration commences in full activity, then a beneficent spirit of warmth pervades the whole frame, and a divine sense of pleasure is all that remains to us of our existence, our whole being seeming dissolved in fleeting vapour. All pains and stiffnesses vanish from the limbs, and we feel light and buoyant as feathers. All bodily pain, be it what it may, disappears in these baths; of headache, toothache, cramps, convulsions in the limbs or face, gout, or rheumatism, there remains not a trace." Another writer says, that *public baths* are more numerous at Moscow than the bagnios at Constantinople. And the author of an article on this subject in the Encyclopædia Britannica, after quoting the preceding passage, says, "One of the greatest improvements of modern times would be the general introduction of the Russian Bath into the towns of Great Britain;" and further, that "the vapour bath is infinitely superior to the warm bath, and, as a medical agent, can scarcely be overpraised." But the pure Roman Bath for which I plead would be free from all the distressing inconveniences mentioned above, would be followed by the same benefits in at least an equal degree, and would be in every respect far better.

Another account still more particular and instructive is furnished us by Dr Granville in his "Travels in Russia." After describing the animated scene presented to a stranger at the usual hour of bathing,—the large courtyard in front filled with carriages and sledges, and crowds of men and women of all classes, some attended by servants, all alike eager to enjoy the healthy and delightful pleasures of the bath, he says, "I was shown into a lofty room 10 feet long by 6 feet wide, the temperature of which varied from 90 to 100 degrees. The inner room is about 75 feet long by 33 feet wide, and of considerable height. The stove here is heated from below, and near it is a door leading into a chamber filled with stones and iron shot kept constantly heated. From the wall, at the height of 10 feet, a brass tube projects, terminating in a round, hollow, and flat rose, pierced with many holes, through which, by the turning of a small cock, water, either hot or cold, may be showered instantaneously, and with considerable force. This room is seldom less than 120 degrees, and frequently from 132 to 140 degrees of temperature. The heat, which at first appeared excessive, becomes gradually more tolerable; nay, one soon gets anxious to experience a little increase. The atmosphere is at this time generally clear. The attendant now approaches to feel the state of the skin, and finding it not quite overspread with perspiration, opens the front door of the stove and throws in a bucket-full of water. Volumes of steam instantly pour forth from it into the room, and a thick fog pervades every part. The body breaks forth into a deluge of perspiration, and all this imparts such a general sensation of comfort as I can scarcely describe. . . . . . In order to excite more perspiration, some bring with them twigs of birch, with leaves on them, with which the attendant gently whips the back of the person bathing. Under this discipline, which the common people inflict very frequently on themselves, the skin becomes of a crimson colour, and perspiration runs out at every pore in such profusion, that none would credit it without actual experiment. It is in this state that many of the Russians have cold water thrown over them, or roll themselves in the snow, or plunge into the nearest half-frozen canal. . . . The physical effects of this bath are highly favourable to the constitution. Judging by my own feelings, I should be inclined to place it above every form of bath in general use; and I think I am indebted to it for the removal of severe rheumatic pains which, before, nothing seemed to alleviate. A Russian is apt to think that almost every disorder to which he is necessarily liable from the severity of the climate may be removed by the bath, and he flies to it on all occasions when ailing. On two other occasions I went to this establishment in St Petersburg, with every symptom of an approaching feverish cold, and returned quite well, and continued so."

As confirmatory of these remarkable and striking results, Atkinson, in his recent book of Travels in Siberia, whither the bath has penetrated also, gives a cure for influenza which he underwent while there, which was quite effectual. He says, " I was boiled alive, then skinned with a birch rod till I presented the appearance of a raw beef-steak, and was finally sluiced with icy-cold water." And Burnes, in his Travels in Bokhara, in which country, and even also among the Tartars in Central Asia, this bath is to be found, says, in reference to the shampooing part of the process,—" You are laid out at full length, rubbed with a hair-brush, scrubbed, buffeted, and kicked; but it is all very refreshing."

Without commending or adopting what appears needlessly violent in the above descriptions, no benevolent and patriotic person can fail to concur in the aspirations and wishes of another eminent traveller in the same regions, namely, the famous Dr E. D. Clarke, who, in his Travels in Europe, says, " Eminent physicians have endeavoured to draw the attention of the English Government to the importance of public baths, and of countenancing their use by every aid of example and encouragement. While we wonder at their prevalence among the eastern and northern nations, may we not lament that they are so little known in our own country? We might, perhaps, find reason to allow that erysipelas, surfeit, rheumatism, colds, and many other evils, especially cutaneous and nervous disorders, would be alleviated, if not prevented, by a proper attention to bathing. The inhabitants of countries where the bath is constantly used, have recourse to it in the full confidence of being able to remove such complaints, and they are rarely disappointed. In England, baths are considered only as articles of luxury. Yet, throughout the vast empire of Russia, through all Finland, Lapland, Sweden, and Norway, there is no cottage so poor, no hut so destitute, but it possesses its vapour bath; whither all the family resort, every Saturday at least, and every day in case of sickness. If some patriotic individual would endeavour to establish throughout Great Britain the use of warm and vapour baths, the inconveniences of our climate might be done away. Perhaps at a future period, donations for public baths may become as frequent as the voluntary subscriptions whereby hospitals are maintained; a grateful people may commemorate the service they have rendered to society by annual contributions for their support. But when we recollect that the illustrious Bacon in vain lamented the disuse of baths, we have little reason to indulge the expectation. At the same time, an additional testimony to their salutary effects, in affording longevity and vigorous health to a people otherwise liable to mortal diseases, from their rigorous climate and unwholesome diet, may conduce towards their introduction. Among the ancients, baths were considered as institutions founded in absolute necessity, and unavoidably due to decency and cleanliness.

Rome, under her emperors, numbered nearly a thousand such buildings, and these, besides their utility, were regarded as masterpieces of architectural skill and of sumptuous decoration. In Russia they have only vapour baths, and these are for the most part in wretched wooden hovels. If wood be deficient, they are formed of mud, or scooped in the banks of rivers or lakes; but in the palaces of the nobles, however they may vary in the splendour of their materials, the plan of their construction is always the same."

To proceed from the Russian to the Moorish Bath, descriptions have been given of the latter both by Mr Urquhart and by Mr Jefferis of the New College, London. Mr Jefferis's account is by far the most animated. He says, "A bath in Bond Street is pleasant enough: nothing seems wanting to luxurious enjoyment. He who thinks so, however, has not tried a Moorish Bath at Tunis. Let me picture one, and the sanitary process going on within. The day was hot; the narrow streets were burning in the glare of noon. The prospect of a hot bath was not very inviting, but I entered. On a raised platform all round the apartment, lay several bathers in a state of profound repose, looking like mummies. The floor was sloppy with condensed steam. In the third apartment the thermometer stood at 160 degrees. It was filled with vapour, which curled round in little eddies at the dome-like roof. I perspired profusely at every pore, and soon my turn came. The bath attendant conducted me, with an encouraging smile, into the second room. He was a strange-looking fellow, his skin as smooth as a chestnut, and his figure plump. He rubbed and pinched every part of my body, lathered me from head to foot, and then proceeded to scrub me with a huge glove, by which the crust of the body was peeled off in flakes. Having been well soused in cold water, the bracing effect of which was highly agreeable, I was softly wiped and dried, and, this done, wrapped up from head to foot in a succession of soft towels, and led to the outer apartment, where I was carefully covered up on a divan. And now commenced a state which many have attempted to describe, but have attempted—only to fail. It was ecstatic enjoyment; it was Elysium. Nothing seemed wanting to perfect bliss but the thought that I could not lie there for ever."

To come next to the Egyptian Bath. It is an offshoot from the Turkish Bath, and the following sketch of it is given by Bayle St John in his "Village Life in Egypt:"—"We went to the bath to be sweated and scraped, and rubbed and lathered and soused, in company with the respectabilities of Siout—brown-skinned, hairy, rotund gentlemen, who submitted to the operation with a gravity and sedateness at once admirable and ludicrous. As they lay like porpoises about the slushed benches, it was evident they felt what important people they were—citizens of a place which possessed a real bath, as not all the race of Pharaoh bathe. From Cairo to Siout we had not found one of these luxurious establishments. In the antechamber, one old gentleman, who had doubtless been soaking for hours, came and sat down, wrapped in a sheet, opposite to us."
And Sir Arthur Clarke says, with a fuller knowledge of the facts, "Bathing, with the Egyptians, as well as with

the Russians, makes a part of their daily wants, and is used as a luxury. In every town and village there is a bath, the use of which has a powerful influence on the health of the people, by removing the causes of those complaints which would seriously afflict them without such prevention, in a climate where perspiration is so copious, and where frequent ablution is so necessary. By this means they avoid a number of cutaneous diseases, as well as rheumatism, catarrhs, and fevers."

In Bonar and M'Cheyne's "Narrative of a Mission" to the lands of the Bible, there is mention made of a visit to the *hamâms* of Alexandria. "The attendant conducted us into an inner apartment, the atmosphere of which we could scarcely breathe at first, on account of the heat and vapour. He laid us down on our back, washed us with soap, and poured hot water over our heads. All this was done by an Egyptian almost naked, armed with a rough glove of camel's hair. We were then led to one of the side baths, where the hot water was allowed to pour upon us. The pores being abundantly opened under the operation of so many causes, we were then shampooed and scraped. Lastly, we were offered coffee and a glass of sherbet, after which we were allowed to dress and come away, not a little amused as well as refreshed. The custom of passing from the bath to the dressing-room, during which the feet might easily be soiled, reminded us of the true rendering of the precious words of our Lord, 'He that has been in the bath needeth not, save to wash his feet, but *is clean every whit.*'"

At Cairo, where the bathers are very numerous, the

hot room is, according to Dr Madden, "so filled with vapour that no object can be distinctly seen at the distance of a few feet from the bather. He sees forms of persons near him, or passing to and fro, as if they were in a thick mist." Archdeacon Goold says, "The sensation produced by the soaping was most delicious. I felt as if I had been lathered by thick Devonshire cream, the soap was so soft, so soothing, so balmy." Thackeray, in his "Journey from Cornhill to Cairo," says, "The after-bath state is the most delightful condition of laziness I ever knew, and I tried it wherever we went afterwards on our little tour." And in the "Cornhill Magazine" the effects of the bath are thus described:— "The blood circulates freely, the chest dilates, the fresh air comes charged with vitality, the wretched find life tolerable, and the aged cast off for a moment the burden of years."

Making our way along the Levant, we find the bath in places in immediate propinquity to its original source, in Smyrna, at Tiberias, and at Deir-el-Kamar, in the Lebanon, Syria. Bonar and M'Cheyne describe a visit to the "Hamam Taberiah," or hot baths of Tiberias, which is supposed to occupy the site of a fenced city called "Hammath," mentioned by Joshua. This bath is supplied with hot water from hot springs. "We," they say, "enjoyed the luxury of a free and copious perspiration, and were afterwards refreshed with watermelons and coffee." At Deir-el-Kamar, the bath, as eloquently described by the late Eliot Warburton, appears to be particularly magnificent: "A Turkish Bath is a very complicated business; but, as it is one of the greatest luxuries of the East, and indeed almost a necessary of life, it is fit to give some description of it; this will equally apply to all, from Cairo to Constantinople. We were first conducted into a beautiful pavilion of pale-coloured marble, in the centre of which crystal streams leaped into an alabaster basin from four fountains. Vases of fresh flowers were tastefully arranged round the carved edges of the basin. A ceiling of soft green and purple porcelain reflected the only light that fell upon this pleasant place. As soon as we laid aside our clothes, attendants brought long napkins of the softest and whitest linen, which were wreathed into turbans and togas around us; then, placing our feet in wooden pattens, inlaid with mother of pearl, we walked on marble floors through several chambers and passages of gradually-increasing heat, until we reached a vaulted apartment, from whose marble sides gushed four fountains of hot water. Here cushions were laid for us, and we were served with pipes, and narghilehs, and iced sherbets; thence we were conducted into the innermost and warmest apartment, where we sat down on marble stools, close to fountains of almost boiling water. This was poured over us from silver cups, and we were then covered with a rich foam of scented soap, applied with the silken fibres of the palm (tree), then bathed again with warm water, and shampooed, in which process the whole skin seemed to peel off, and every joint was made to crack. Then we were again lathered, and again soused, and found our skins as soft as that of a little child. We now left the warmest room, and were met at the door by a slave with bundles of exquisitely soft warm linen, in which we were again shawled, turbaned, and kilted; and so we passed out into the cool fountain-chamber, where another change of linen awaited us. It was a sudden and pleasant alteration from burning suns, and craggy roads, and sweltering horses, to find ourselves reclining on silken couches, in the shaded niche of an arched window,—through which cool breezes, filled with orange perfumes, breathed gently over us. The sensation of repose after a Turkish Bath is at all times delicious; but here it was heightened by every appliance that could win the entranced senses to enjoyment, without disturbing their repose."

There is a curious passage in the Mishna, or oral law of the Jews, from which some might be disposed to deduce the conclusion that the principle of the bath of which we speak was known among them. Speaking of the service of the Temple and its arrangements, it says, " Lastly was the bath-room. There the priests descended to the bath (beneath the platform of the sanctuary). There likewise fire was piled up, at which the priest, after he had washed his body, might be made warm, and, ascending from the bath, might wipe himself; and it was named *the room of the heat*, and it opened towards *the greater place of the heat*, as it was called. It was called the House of Washing. No one entered the court for the ministry to be performed before he had washed."

Coming to the capital of Turkey itself, we find a very intelligent account of what must be a real Turkish Bath in Gautier's " Constantinople of To-day." On his entrance into it, after noticing in the beds, smoking, drinking coffee, or sleeping wrapped to the chin, the bathers—that is, those who have bathed—awaiting the moment when perspiration shall have ceased sufficiently to permit them to dress, he observes, " These baths differ very much from our vapour baths. A fire burns continually beneath their slabs of marble; and the water which is spread above them is volatilized in clouds of white steam, instead of issuing from a boiler in jets. These are, in a manner, *air baths*, and the extreme heat induces the most profuse perspiration. The tellak, with a gauntlet of camel's skin, begins to curry the bather, dashes over him basins of warm water, and finally polishes him off with the naked hand, causing grey rolls to peel from the skin, in a manner astounding to a European convinced of the cleanliness of his own person. My tellak was a young Macedonian of fifteen or sixteen years of age, whose skin, softened by constant immersion, had acquired a fineness of polish almost inconceivable. When I at length arose and went forth, I felt so light, so supple, so relieved from all sense of fatigue, that I appeared to be walking on air."

Savary, a French author of the last century, gives a description which has been much quoted and admired, and, for its present truth and applicability, might appear to have been written yesterday. A few sentences will suffice :—" During the operation, he detaches from the body of the patient, which is running with perspiration, a sort of small scales, and removes the imperceptible impurities that stop the pores, and the skin becomes soft and smooth like satin. Coming out of a bath filled with hot and moist vapour, where the perspiration gushed from every limb, and transported into a spacious apartment, open to the external air, the breast dilates, and you breathe with voluptuousness. Well kneaded, and, as it were, regenerated, the blood circulates with freedom, and you feel as if disengaged from an enormous weight, together with a suppleness and lightness to which you have hitherto been a stranger. A lively sentiment of existence diffuses itself to the very extremities of the body. While it is lost in delicate sensations, the soul sympathises with the delight, and enjoys the most agreeable ideas. The imagination, wandering over the universe which itself embellishes. sees on every side the most enchanting picture, and everywhere the image of happiness. If life be only a succession of ideas, the vigour, the rapidity with which the mind runs over the extended chain, would induce a belief that in the two hours of that delicious calm that succeeds the bath, one has lived a number of years. Such, sir, are these baths, the use of which was so strongly recommended by the ancients, and the pleasures of which the Turks still enjoy. Here they prevent or exterminate rheumatisms, catarrhs, and those diseases of skin which the want of perspiration occasions. Here they rid themselves of those uncomfortable sensations so common among other nations, who have not the same regard to cleanliness."

A few words from Lady Mary Wortley Montagu, and an American traveller, will complete the picture. The narrative of the former, written early in the last century, is interesting as affording us the rare opportunity of seeing the interior of a ladies' bath in Turkey. The bath at St Sophia, in Constantinople, was already full of women, and they received her, she says, " with politeness, and without the least surprise or impertinent curiosity. I believe that there were two hundred women, and yet there were none of those disdainful smiles and satirical whispers that never fail in our assemblies. The first sofas were covered with cushions and rich carpets, on which sat the ladies ; and on the second, their slaves behind them, but without any distinction of rank by their dress, all being in the state of nature—that is, in plain English, stark naked. They walked and moved with the same majestic grace which Milton describes our general mother with. There were many amongst them as exactly proportioned as ever any goddess was drawn by the pencil of a Guido or Titian—and most of their skins shiningly white, only adorned by their beautiful hair, perfectly representing the figures of the Graces. I was here convinced of the truth of a reflection I have often made, that if it were the fashion to go naked, the face would be hardly observed. I perceived that the ladies of the most delicate skins and finest shapes had the greatest share of my admiration, though their faces were sometimes less beautiful than those of their companions. The baths, fountains, and pavements are all of white marble, the roofs gilt, and the walls covered with Japan china. Adjoining to them are two rooms. the uppermost of which is divided into a sofa, and in the four corners are falls of water from the very roof. from shell to shell of white marble, to the lower end of the room, where it falls into a large basin surrounded with pipes that throw up the water as high as the roof. '

Having been importuned by her companions to take a bath, she was at last forced to open her dress and show them her stays, which, she says, "satisfied them very well; for I saw they believed I was locked up in that machine, and that it was not in my own power to open it, which contrivance they attributed to my husband." Further, she says, they were occupied, some in conversation, some working, others drinking sherbet, and many lying negligently, getting their hair braided by their pretty slave girls. They generally take this diversion once a week, and stay there at least four or five hours without getting cold, by immediately coming out of the hot bath into the cold room, which was very surprising to me."

Last of all, Mr N. P. Willis very graphically says, "We were led into the first bath, a small room in which the heat, for the first breath or two, seemed rather oppressive. After a half hour, the atmosphere, so warm when we entered, began to feel chilly, and then we were led into the grand bath. The heat here seemed to me for a moment intolerable. The floor was hot, and the air so moist with the suffocating vapour, as to rest like mist upon the skin. It was a spacious and vaulted room, with fifty small square windows in the dome. In the centre was a broad platform, on which the bather is rubbed and shampooed. occupied just then by two or three dark-skinned Turks, lying on their backs, with their eyes shut, dreaming. if one might judge by their countenances, of Paradise." After the operation of the bath, he says, "In a few minutes we began to feel a delightful glow in our veins, and on returning to the dressing-room, my sensations were indescribably agreeable—absolute repose of body, and a calm tranquillity of mind, equally unusual and pleasurable. It seemed as if pleasure was breathing from every pore of my cleansed and softened skin. I would willingly have passed the remainder of the day upon that luxurious couch. One could hardly fail to grow a poet, even with this habit of Eastern luxury alone. If I am to conceive a romance, or to indite an epithalamium, send me to the bath on a day of idleness, and, covering me up with their snowy and perfumed napkins, leave me till sunset!"

Our survey would not be complete without a few concluding words on the bath in its best estate—namely, as it existed in practice among the Romans. From the foregoing quotations, which ought to be well pondered, as furnishing us with facts all tending to prove the inestimable advantages of the bath, and which, besides, are of varied literary and social interest, it has plainly appeared that the Turkish Bath,—or the Roman Bath, as practised by the Turks,—is an air bath, and, though numid, a hot-air bath, and not a mere vapour bath; and that, though there is a varying and uncertain amount of vapour in the air of the bath, according to their practice, yet these Eastern Hamâms and the Russian Baths, and all the others of which I have spoken, owe their superiority to the pure vapour bath to the fact of the air not being *saturated* with vapour, and therefore admitting of evaporation. But I have now to adduce my authorities upon the pure hot-air Roman Bath.

Authorities tell us that the great hall of the Roman

Bath was appropriately ornamented with the statues of Hercules, the god of strength; Hygeia, the goddess of health; and Æsculapius, the god of medicine. Pliny tells us that an incredible number of baths were built up and down the city, many of them erected by the emperors with amazing magnificence. Gibbon says, " The stupendous aqueducts, so justly celebrated by the praises of Augustus himself, replenished the thermæ, or baths, which had been constructed in every part of the city with imperial magnificence. The baths of Antoninus Caracalla, which were open at stated times for the indiscriminate service of the people and senators, contained above sixteen hundred seats of marble; and more than three thousand were reckoned in the baths of Diocletian. The walls of the lofty apartments were covered with curious mosaics, that imitated the art of the pencil in the elegance of design and variety of colours."

Celsus, as might be expected, speaks of the bath in its application to the purposes of health, and in so doing clearly enough expounds its principle. Treating of perspiration he says, " It may be procured in two ways—either by dry heat, or a bath. A dry heat is raised by hot sand, the laconicum (part of the sudatorium) and clibanum (a common stove), and some natural sweating places, where a hot vapour exhaling out of the earth is enclosed by a building, as there is at Baiæ, among the myrtle groves. These kinds are useful wherever an unnatural humour offends, and is to be dissipated. Also some diseases of the nerves are best cured by this method. The use of the bath is twofold. For, sometimes, after the removal of fevers, it is a proper intro-

G

duction to a fuller diet and stronger wine, for the recovery of the health ; sometimes it removes the fever itself. And it is generally used when it is expedient to relax the surface of the skin, and solicit the evacuation of the corrupt humour, and change the habit of the body. A valetudinary man on going to the bath, if he should feel his temples bound and his skin dry on reaching it, had better not take the bath to-day, but return home, as it would be hurtful. And regard must be had to his strength in the bath ; he must not be allowed to faint by the heat, but must be speedily removed, and carefully wrapped in clothes, lest any cold get to him, and there also he must sweat before he take any food."

Turning next to Leo Africanus, we find him saying, " When any one is to be bathed, they lay him along the ground, anointing him with certain unguents, and with certain instruments clearing away his filth."

Pliny, in his Letters, speaks most affectionately of the bath ; as, for instance, " the sunny bath" at Como, " the pretty bath" at Narnia. He describes the one erected at his private villa at Laurentium, its furnace, sweating-room, perfuming-room, and cooling-room; and says, " When the baths are ready, which in winter is about three o'clock, and in summer about two, he undresses himself, and if there happens to be no wind, he walks for some time in the sun. After this he plays a considerable time at tennis, for, by this sort of exercise too, he combats the effects of old age. When he has bathed, he throws himself upon his couch till supper time."

Again, Seneca says in one of his letters, " As yet I have confined my remarks to private baths only. What shall I say when I come to our public baths ? What a profusion of statues ! What a number of columns do I see supporting nothing, but placed as ornaments, merely on account of the expense ! We are come to that pitch of luxury that we disdain to tread upon anything but precious stones. Formerly, baths were few in number, and not much ornamented; the water did not pour down in drops like a shower, nor did it run always fresh and clear as from a hot spring ; and the heat was not like that of a furnace. Many ridicule the simplicity of Scipio. Unhappy man ! The water he washed in was not clear and transparent ! This, however, concerned him but little ; he came to the bath to refresh himself after his labour, not to wash away the perfumes of a pomatumed body."

One of the earliest advocates of the revival of the Roman Bath in this country was Count Rumford, and I shall conclude this chapter by quoting his remarkable testimony in its favour :—" Were the general and constant use of the warm bath by persons in health a new thing, I should have many scruples in recommending it to the public, whatever my private opinion of its salubrity might be. But so many nations have practised it for ages, and there are so many who now practise it, and, what is very remarkable, one—the Russian—which inhabits the coldest part of the globe, that there cannot possibly be the smallest reason to doubt of its beneficial effects. With regard to the pleasant effects that result from the use of the warm bath, there never has been any difference of opinion. But still I am quite certain that the true luxury of warm bathing is not understood in this country, and till the construction of our baths is totally changed, and a different manner of using them adopted, we never can enjoy a warm bath as it ought to be enjoyed. As we must allow that in most cases, and particularly in a matter of this kind, it is much more wise and prudent to adopt those arrangements and improvements which have been the result of the experience of ages, than to sit down and invent anything new, I think we cannot do better than rebuild some of the baths which were left us by the Romans. They most certainly understood warm bathing as well as any nation ever did, and if there be anything in our climate which renders any deviations necessary from the manner commonly practised in constructing baths in warmer countries, there is no doubt that those luxurious foreigners, who had possession of this island for so many years, must have found them out. The plans they have left us may therefore be adopted with safety as models for our imitation. I cannot help wishing that the inhabitants of this island, and all mankind, might enjoy all the innocent luxuries and comforts that are within their reach. I am even jealous of the poor Russian peasant; and when I see him enjoying the highest degree of delight and satisfaction in the rude cave which he calls a warm bath, I greatly lament that so useful and so delightful an enjoyment should be totally unknown to so great a portion of the human species. When I meditate profoundly on these subjects, it is quite impossible for me not to feel my bosom warmed with the most enthusiastic zeal for the diffusion of that knowledge which contributes to the comforts and enjoyments of life."

Such are the fruits of our survey. Such and so varied are the benefits which different nations feel that they derive from this species of the warm bath. Farther on I shall consider and explain its beneficial effects more fully and distinctly. But one thing must surely strike the attention of every reader of the preceding survey, especially in drawing near to its close, and that is a possible explanation of the great contrast between the ancient Romans and the modern Italians. And the inquiry suggests itself, May not the immense difference between them—the indomitable energy and conquering valour of the old Romans, and the comparative degeneracy of their descendants—be owing, in a considerable measure, to the extensive use of this bath among the former, and to its comparative disuse and decay among the latter?

CHAPTER V.

PHYSIOLOGICAL ACTION OF THE BATH.

IN order to understand the *rationale*, or principle, of the Turkish Bath, it will be necessary that I should give a brief description of the anatomical structure of the skin, and of the manner in which it is affected or acted upon by the external agency of heat. Physiology is the science of living things, and comprehends the phenomena of life and natural organisation. The more correct term of " Biology" has been employed by the Germans, and, as opposed to the wider signification—nearly synonymous with Natural Philosophy—formerly given to the word " Physiology," is now employed to designate the science in its more restricted application to the functions of the living organism, and the powers by which these functions are exercised. In proceeding to speak of the philosophy of the bath, or the physiological action of the hot-air bath upon the human body and the multitudinous organs of which it is composed, it is proper to advert first to the skin and its functional structure, for it is upon the skin that its influence is more immediately and sensibly exerted. Thus will the *modus operandi* of the bath be best understood.

The surface of the skin, or epithelium, is formed of

laminæ, or scales, which are continually being detached by the friction of the clothes, and as continually reformed by condensation of the subjacent cell-membrane. These laminæ, or scales, are thick and adherent, in proportion to the pressure to which they are subjected, as may be seen on the hands of artisans and the feet of pedestrians. Now this admirable provision is evidently designed for the protection of the delicate and sensitive skin beneath it. Through the epithelium two kinds of ducts pass—the ducts of the sudoriferous and of the sebiparous glands, the orifices of which constitute the pores of the skin. The skin is penetrated besides by the innumerable hairs dispersed over the surface of the body, formed by the secretion from the hair glands. The cutis, or true skin, covered by the epithelium, comprises the nerve papillæ, constituting the organ of touch, the sudoriferous or perspiratory glands, with their several capillary or hair-like arteries, veins, and absorbent vessels, united into one strong tough membrane by the areolar tissue. Beneath the cuticle and cutis are the pigmentary glands, which give the colour and complexion to the body, varying in intensity of shade from the fairest to the darkest races of the human family.*

From this brief anatomical sketch, the great importance of the healthy condition of the epithelium or cuticle, and the cutis or true skin beneath, to the wellbeing of the whole body, is manifest. If the pores which pass through the cuticle are obstructed, the cutis inevitably becomes diseased. If the secretion of the perspiratory glands is arrested, the blood becomes con-

* See Dublin University Magazine, 1861.

taminated, and diseases of various kinds are produced in the internal organs ; so that it may be asserted, without hesitation, that the wellbeing of the whole body depends upon the healthy condition of this important integument. For if the dead scales of the epidermis, or outer skin, which are constantly being destroyed and renewed by the action of the true skin beneath, are allowed to accumulate, like moss upon the bark of trees, the effect upon the bodily health must necessarily be pernicious. No mere sponging with cold water will remove them. These dead scales becoming saturated with perspiration and in a state of unwholesome ferment, the pores beneath are clogged up altogether, whereby the necessary elimination of poisonous matters from the body is prevented, and further, the absorbent vessels suck up the impure and decomposing perspiratory matter wherewith the skin is covered. Thus the blood is rendered impure, the whole system is contaminated, and the foundation is laid for many deep-seated and destructive diseases, which are actually generated by the suspension of the proper action of the skin. In this we have an explanation of the origin of many diseases in the internal organs. And what are pimples, boils, pustules, carbuncles, &c., in the skin, but just Nature's efforts to relieve the blood of the impurities which cannot get properly escaping through the pores ?

The vitally important function of the skin in the human economy will be apparent to every one who considers for a moment that it is a most powerful eliminator of the poisons called Hydro-carbons from the blood. Its extensive surface is thickly studded with vessels, sudoriferous tubes, sebaceous glands, and lymphatics. It has been estimated that there are actually no less than seven millions of pores opening upon its surface, and that the ducts of its glands, if placed in a straight line, would extend over from 27 to 30 miles in length. Through this extensive system of bodily drainage, there ought daily to distil, on an average, more than 21 ounces of fluid, holding in solution, according to an eminent chemist, from a half to one and a-half per cent. of solid matters, consisting of a compound portion in a state of incipient decomposition, and saline matters of various kinds. Besides this eliminating function of the skin, it is possessed of a power similar to that of the lungs, of absorbing oxygen and setting free carbonic acid gas. So that if the body were coated with an impermeable varnish, death would speedily be the result, from reduction of the animal temperature and non-aëration of the blood. In proof of the injurious effect of a stoppage of the functions of the skin, we have only to refer to a fact recorded in the history of the dark ages, during the reign of one of the popes. When Leo X. was raised to the pontificate, a child was gilded all over at Florence, to represent the genius of the golden age. And what was the consequence? The speedy death of the poor boy. And why? Because the skin could not breathe and inhale the necessary oxygen from the air. Such was the result of complete closure of the pores of the skin; and what follows partial closure of the same organ to any considerable extent? Disease in a thousand forms; but most commonly and most certainly that which gives annually in this country more victims to the grave than all other diseases put together. *scrofula*, including among its subtle and deadly forms, *consumption*. Effete matter seeks for oxygen ; and if it cannot find it at the skin, it flies to the lungs, and ultimately destroys the organ by which it sought an exit from the body. But, by means of the hot-air bath, such scrofulous deposits, which would otherwise fly to the lungs, are drained off by the skin when it is stimulated to activity.

The unfortunate result of the stoppage of the pores of the skin in the case of the youth, as mentioned above, shows the gross ignorance that existed in that dark period of the world's history of the anatomy of the skin, and its vital importance as an organ which plays so mighty a part in maintaining the health, vigour, and comfort of the living body. The great blood-purifier of the system is the lungs, and the only other organ that at all approaches to them in importance in this respect is the skin, for the simple reason that a part of its function is the absorption of oxygen, which is the breath of life. That the skin is a *Breather*,—a *Respiratory Organ*, to a certain extent, and thus a *Maintainer of animal heat*, is undoubted. Some animals have no lungs, and breathe entirely through the skin. Others have a portion of the skin modified into gills, or rudimentary lungs. And in the higher animals, though the lungs have become especially devoted to this function, still the skin retains it to such an extent that to interfere with it, or obstruct it, is dangerous—totally to arrest it, is fatal. It exhales carbonic acid, and it inhales oxygen, and this may easily be proved by holding the hand in a vessel of oxygen, when that gas will shortly be absorbed and replaced by carbonic acid.

What is still more remarkable is, that the functions which the skin performs are frequently vicarious in their nature, that is, in respect of the secretions of internal organs, whose functions have become impaired through organic disease. In this view, the great importance of the excretions from the skin is evident, as a means of purifying the blood, and preserving the health. Not merely from the anatomy and physiology of the skin, but especially from the consideration of the vicariousness of its function when occasion requires, its extreme pathological importance is at once manifest, and the value of a clean and sound skin as preservative of health, preventive of some diseases, and alleviative of others, is at once apparent. Thus the skin may, and actually does, frequently become vicarious in its action in respect of the lungs. In this way it becomes highly useful to the economy, in relieving the lungs when overtaxed or diseased. And hence the great value of the hot-air bath in cases of consumption and all diseases of the chest. This intimate sympathy between the skin and the lungs may be apparent to any one in health, by simply observing, when walking fast, how much more easily he gets along his journey after he has broken out into a perspiration; or the rider may observe how his horse freshens up under the same condition. Exercise is, of course, the legitimate way of increasing the action of the lungs as well as of the skin, and of exciting them to healthy activity. But in the case of patients, who complain of weakness of the chest, or other disorders of the respiratory organs, and who are unable to take the necessary and salutary amount of exercise, the next best thing we can do for them is to set free the pores of the skin, to stimulate and call into active play that set of organs which can act vicariously of the lungs, and which are to be found in the skin. And this can only be done by means of the hot-air bath, which, besides its eliminating power, combines the luxury of the warm bath with the genial and refreshing influence of the tepid, the bracing effect of the cold douche, and, better than all, the absorption of oxygen. Herein lies the philosophy of the Bath, and in further illustration of its physiological action, I desire to enrich my pages by the following quotation from Dr Carpenter, the most popular writer of the day on the subject of human physiology. He says :—" With regard to the functions of the skin taken altogether as a channel for the elimination of morbific matters from the blood, it is probable that they have been much under-rated, and that much more use might be made of it in the treatment of diseases—especially of such as depend upon the presence of some morbific matter in the circulating current—than is commonly thought advisable ; we see that nature frequently uses it for this purpose, a copious perspiration being often the turning-point or crisis of febrile diseases, removing the cause of the malady from the blood, and allowing the restorative powers free play."

The skin is the safety-valve for all the inner vital organs. It is furnished, as we have seen, with a vast system of drainage of many miles in extent, in its millions of pores. To cleanse the body properly, it is necessary, first of all, to empty, to purify, and flush these drain-pipes with which its surface is so profusely covered. For the maintenance of health, this system of sewerage, as supplementary to the lungs, requires to be kept in regular working order and in constant use for the purifying of the body. It is the only organ specially consigned to man's own care, and placed under his own observation, and yet it is the one most neglected. To keep it in natural healthy action, exercise, inducing perspiration, is requisite, or, in lieu of this, the hot-air bath. The hard-working man's shirt, for example, smells abominably at the end of a week, but the drainage of his body is in good working order. He labours and perspires. "In the sweat of his brow he eats his bread." Consequently he is strong and healthy, and his system is really cleaner within than that of those who enjoy change of raiment every day. Many an individual, on the other hand, whose occupation prevents him from taking sufficient bodily exercise, and who is consequently weak and delicate, may look externally cleaner than the working man, and yet be really less so internally, from the impure matters retained within his system, and not thrown off, as they ought to be, by perspiration.

A writer in the "Cornhill Magazine" for March 1861, in an article on Turkish Baths, very justly says on this point, "The gentleman who takes his trough bath is internally dirtier than the working man who has been digging a ditch all day in the hot sun." And it is very obvious why it should be so. The ditcher has been perspiring profusely. The pores, or drain-pipes, in his body have been as it were flushed, and the skin is in a condition fully to perform its proper functions. His system, therefore, succeeds in energetically casting forth and ejecting, through the pores of the skin, all the used-up, effete, and superfluous materials. Thus the labouring man is in perfect health. The condition of his body is in many respects enviable. His appetite is good. " His sleep," as Solomon says, " is sweet, whether he eat little or much." But as we cannot all be labouring men and dig in a ditch, and, from the varied nature of our employments, cannot all secure a copious perspiration, and hence bodily vigour and immunity from disease by the natural methods which the labourer makes use of, we have a most valuable and agreeable substitute placed within our reach in the Turkish Bath. Man was made for an active life in the open air. But our present modes of life are artificial. We labour and work at our various employments, but we do not perspire. We are busy all day, and fatigued at night, but still we do not perspire. We do not eat our bread in the sweat of our brow. But in the artificial hot-air bath we perspire. Here we find a substitute—a succedaneum.

The means of preserving health by maintaining the normal state of the skin, have claimed the attention of civilised nations in all ages, and that in proportion to their civilisation. These means are comprehended under three heads—friction, ablution, and perspiration. The first only removes the laminæ of exfoliated cuticle from the surface ; the second removes obstructions from the pores or orifices of the ducts ; the third, by increasing the secretions of the sudoriferous and other glands, removes obstructions in the ducts themselves and not merely from their mouths, and thus preserves the body in sound health. Now, in its *modus operandi*, the thermal bath precisely imitates natural efforts to expel morbific and diseased matters by perspiration. For example, in the treatment of patients whose constitutions have been weakened by frequent attacks of fever of an aguish kind, the thermal bath has been found very useful. Complaints of this nature are extremely liable to recur for the remainder of life, from any cause which chills the body and checks the perspiration, even long after the removal from the locality where the disease was first caught. For such cases the Turkish Bath is an effectual and specific remedy.

Modern purification has been defined as superficial—ancient, as reaching the blood and all the tissues of which the body is composed. Now, predisposition to disease consists more or less in diminution of the energy of the capillary circulation, in consequence of which the frame becomes less able to resist the injurious impressions of various morbific agents, such as atmospherical vicissitudes, miasmatic emanations, &c. And when we consider the subjects in which scrofulous diseases take place, and the multifarious forms which it assumes. it may be inferred that this predisposition consists in a peculiarly inert and relaxed state of the whole capillary vessels, by which they either lose the power of depositing firm, solid particles in the different tissues, or are liable to deposit particles of albuminous matter, void of organisation, and consequently incapable of forming healthy portions of the animal tissues. This habit or disposition of body is believed to be the internal cause which, with or without the application of exciting external causes, gives rise to all the diseases denominated scrofulous.

I may explain here, that these small capillary vessels, or hair-like tubes, of which I am now speaking, are the terminations of the arteries. They are so minute that they cannot be seen by the naked eye. Each of them has its accompanying vein and nerve. Viewed altogether, they are called by medical men the ultimate tissue.* This tissue pervades every organ, and every part of the frame, so that you cannot insert the point of the finest needle into any part of the body without wounding one or other, perhaps many, of these minute hair-like tubes. Now these small arteries, at their very extreme points, are reflected back upon themselves. Here they lose the character of arteries and become veins; and while the former carry blood and nutrient material to the various parts of the organism, the latter carry off the impure blood back again to the heart. From thence it is propelled to the lungs, where it is purified by the oxygen which we inhale, and, again furnished with nourishing material, proceeds on its course to supply the waste continually going on in the body. It is in this ultimate tissue that disease first invades the system. Now, let it be remembered that this ultimate tissue, with its vast and complicated plexus of veins, nerves, and arteries, extends to every part of the body, but has its principal seat in the skin, where the innumerable extreme points or termini of these mi-

* See Dr Craigie's Practice of Physic.

nute vessels are all concentrated together, and then the great importance of preserving the skin in a healthy condition becomes at once apparent.

I am the more particular on this deeply interesting point, because it is so closely connected with the subject of the hot-air bath, which is calculated to restore the ultimate tissue to its normal state, when, through any cause, its functions have become depraved and deranged, and its energies impaired, and also because it is admirably adapted to maintain this ultimate tissue in sound working order. The subject is of vital importance to our well-being, because, in proportion as we attend to the voice of Nature's laws, in living, eating, and drinking—in cleanliness, air, and exercise,—we experience health of body, together with vigour of mind and elasticity of spirits.

Before I go on to describe the part which heat plays when artificially applied to the ultimate tissue, by means of the hot-air bath, in further illustration of its physiological action, I cannot help giving a single quotation from the valuable work of Dr Johnson, on "Life, Health, and Disease." It is very much to the point, and illustrates with much force and perspicuity what I have now been endeavouring to urge. Speaking of the causes of disease, he says :—" Let us inquire as to what is the actual condition of the solids and fluids of the body in these distressing circumstances of the health. I believe it to consist in sanguineous congestion of the ultimate tissue of all the organs concerned in the nutrition of the body, and, as a necessary consequence, in a deficiency and depravity of all the necessary secretions.

The blood, therefore, when not circulated with sufficient energy through this ultimate tissue, becomes deteriorated in quality, and this, too, precisely where it is of the utmost importance that it should be of the very highest degree of purity." Again he observes, " It is in the ultimate tissue of our organs that all those operations are effected in the blood, on which the nourishment of the body depends; and that this deranged and diseased condition is invariably the result of people's own conduct in errors of diet, drinking, and a host of other causes."

Seeing, then, that it is in this part of the organism that disease first invades the system, the question very naturally suggests itself, What are the best means to avert its farther progress—to nip it, as it were, in the bud? You will be told, perhaps, to take this or that medicine, or to rub some liniment on the affected part. If this fails, as it often does, then probably you will be subjected to what is called " active treatment," such as the abstraction of blood, either local or general, to be followed by blisters and counter-irritants. Should it prove of a very obstinate character, and resist all these strong means, then a course of mercury, combined with opium, must be had recourse to as a forlorn hope. Now, how much better would it be to proceed rationally to work, and, instead of using any of the measures referred to, at once to attack the enemy in his stronghold, and not in some of his advanced outposts merely,—to drive him out by storming his first position in the ultimate tissue? This, I boldly affirm, can only be done by such means as the hot-air bath. Let the preceding observations on the active causes and the real seat of disease be duly considered, and let the condition of the body on the first invasion of disease, or congestion of blood, and consequent languid circulation, and deterioration of its quality, be fully weighed, and it will be perceived at once by every unprejudiced mind, that the Turkish Bath is just the very thing that is wanted, and that, when judiciously administered, it will be found a most powerful and sovereign remedy. After long experience of its virtues, I have no hesitation in asserting, that it is impossible to speak too highly of it as a boon to many individuals, whose activity and usefulness have been destroyed, and whose lives have been rendered burdensome, alike by the disease and the means resorted to for its cure.

We are now prepared for the question—How does it act? By opening the pores, by setting free the accumulated excretions which have clogged and blocked up the countless drain-pipes of the system, and by exciting a brisk, lively, and healthy circulation throughout the whole of the ultimate tissue, and thence throughout the whole body. As Dr John Armstrong, of London, says in one of his lectures on congestive fever, "The hot-air bath does not fatigue ; it will bring pounds of blood to the surface which were suffocating some internal organ ; it will balance the circulation sooner than any other means I know." Such are its truly marvellous and beneficent effects ; such is the physiological action of the Turkish Bath.

But by what means, and by what agency, is this accomplished ? By the agency of HEAT. and more particularly by the application of heated dry air to the whole surface of the body. The application of increased heat to the bodily surface, either with or without medicine, is a system in itself. Moreover, it is a system which offers no opposition to the usual modes of combating disease. Each may be made subservient to the other, where either alone is insufficient to insure success. It is singular that heat, an elementary property so well known and understood, should have been always floating, as it were, before our senses and within our reach, and that, until of late years, it should never have been seized on and brought into useful requisition for the treatment of disease. Heat softens, expands, loosens, and separates all substances, even the hardest. Applied to the body in the shape of heated air, it expands all those small vessels where the extremities of the veins and arteries unite in the skin and throughout the whole animal tissue. All the parts being thus expanded, disease which may have taken hold of any part is weakened and loosened in its hold, its escape or removal from the system is facilitated, and nature is aided in her efforts to throw it off through the various emunctories, and more particularly through the myriads of opened outlets in the skin. Mercury, when taken in large quantities produces the same expansion and loosening of the parts, even to the teeth ; but it is a dangerous and desperate remedy, and terrific in its effects. Heat, on the other hand, as applied through the hot-air bath, is harmless and safe, besides being a far more pleasant, speedy, and effectual remedy. *

* *Vide* Dr Jonathan Green, on the Fumigating Bath.

We see how indispensably necessary heat is to both animal and vegetable life, to fœtation, incubation, and germination, and in every stage of existence. It is so closely connected with vital energy, that its influence is absolutely necessary to animate, invigorate, and beautify the works of creation. It enables plants to grow, put forth their flowers, and mature their fruit. Persons apparently drowned are mainly recovered and resuscitated by the application of heat. While cold nips and benumbs, retards and depresses the functions, and diminishes and contracts the action of the vessels of the skin, the temporary application of heat occasions a renewed energy to all the vital functions. It is in this way that the hot-air bath is so valuable and beneficent in its operation. By gently stimulating the whole frame, it quickens and increases the secretions drawn from the blood for the nourishment of the body, "produces a free and full circulation, removes all obstructions in the vascular system, and puts all the organs into that state of regular, free, and full motion which is essential to health, and also to that delightful repose, accompanied by a consciousness of the power of exertion, which constitutes the highest animal enjoyment of which we are capable."*

But it may be asked, will not the application of a high degree of external heat to the body be dangerous? Will it not elevate the temperament of the body to an undue and dangerous degree? To this I answer, no such danger exists. For, besides those functions of the skin to which I formerly adverted, it performs another equally

* Count Rumford on Bathing.

important, namely, in maintaining a moderate temperature within the body, no matter how great the external heat to which the living body may be exposed. This is effected by means of evaporation from the cutaneous surface, by means of which the body is kept cool, without any appreciable increase of its temperature, even though exposed to a heat sufficient to cook animal food and to boil eggs. Monsieur Chabert, the Fire King, exposed himself with impunity several times a day to a temperature of several hundred degrees in an oven, protecting his feet from the hot bricks of the floor by means of thick cork soles, and actually cooking his own food by means of the high temperature of the oven. The average temperature of the human body is about 97 or 98 degrees, and a thermometer placed under the tongue of man in all countries rises only to this height. Dr Jonathan Green says, that by repeated experiments with a thermometer placed under the tongue, while the patient was exposed to a high degree of heat, he found he could not raise the heat of the body to 99 degrees, for no sooner did perspiration commence than the cooling process also began. To whatever degree of heat or cold the body is exposed, we find that the blood and the internal parts always indicate nearly the same degree of temperature.

To explain this, let me briefly sketch the natural process which takes place when the body is subjected to the influence of the hot-air bath. The seven million of pores are freely opened. The vast net-work of blood-vessels and nerves, 2500 square inches in extent on the skin of an ordinary-sized man, and of the finest conceivable texture, is excited to unwonted activity when brought under the genial influence of this powerful agent and solvent. The body being immersed in an equalized temperature of hot air, every single pore on its whole surface feels its salutary influence. It expands the skin, and insinuates itself among all the interstices of its laminated structure. The perspiratory fluid now excited bursts forth in profuse streams. Nature is now relieved. The circulation is no longer deranged, and the congestion of the internal organs is removed. Irritability is allayed, and a feeling of health and comfort pervades the frame.

Now, when the living body is thus surrounded with heated air, and a free perspiration has been induced, if the heat be increased and continued for a time, the sweating is increased in proportion to the stimulus from the excess of the heat. But the body itself is not thereby unduly heated, nor is it in any way injuriously affected. For the efforts of the system are now brought into play to counteract the effects of accumulation of heat in the body. The refrigerating process is effected by the evaporation of heat from the body in the form of vapour and perspiratory matter. Dr Gairdner says on this subject, " I should not have insisted so much on the effects of evaporation, had not I considered it a material circumstance in examining the effects of hot air on the human body, which, sooner or later, according to the degree of heat it possesses, produces in the manner above mentioned a sweat, and consequently evaporation from every part of the body. Not that the whole of the matter perspired is turned into vapour. It is only such a portion of it as can readily absorb the necessary quantity of heat from the body and the external air, which will be in proportion to the degree of heat they possess." And thus it is that the human frame is enabled to maintain its natural temperature, notwithstanding the amount of heat to which it may be exposed.

Thus, then, we have the physiological action of the hot-air bath fully explained. In subsequent chapters I shall show its varied and beneficent uses, and its application to purposes of health. I have already shown that, physiologically considered, it is just a dose of heat—a dose of heat in its most agreeable form, capable of being most pleasantly administered, free from all the permanent ill effects of other stimulants, and a dose that may be safely repeated until its ameliorating, healing, and strengthening effects have become fully established. And after this demonstration of the salutary and beneficent action upon the functions of the skin of that institution preserved for us in Europe by the Turks, let us, as Leigh Hunt says, "import from them not only figs, but a fine state of the pores." And let us hope that another Isaac Watts will arise to sing in moral songs the praises of the bath, and to describe the busy bather, and how he improves each sweating hour, and gathers oxygen all the day from every opening pore. For surely it is not too much to expect, in this age of progress and enlightenment, that such a powerful preservative and prophylactic against disease should find universal favour and general adoption, especially when both the negative and the positive benefits derivable from its use are so momentous.

CHAPTER VI.

USES OF THE BATH.

THE object of all bathing is to free the skin from the accumulated deposits of insensible perspiration, and thus to promote the healthy action of the whole system. The object of the Turkish Bath is to purify, refresh, and renovate the system by the most thorough and complete mode of external ablution known to man. In the East, no source of comfort and enjoyment is deemed more essential to existence than this bath. There it is associated with all other pleasures. Our own countrymen, in passing to and from India, eagerly pay a visit to the *Hamâms* of Cairo and Alexandria; while to the traveller arriving in Constantinople, the anticipated luxury and enjoyment of the bath is second only in his mind to the impressions produced by the magnificent entrance into that city. In the East, much more attention is paid to the skin, to keep it in a state of healthy natural action, than among us. Such was the case also among the ancient Romans, whose baths, as we have already shown, were conducted very much on the same principle and plan as the modern Turkish Bath. The most famous physicians of ancient times, such as Hippocrates and Celsus, recommended their use, and they were recommended by Christian and pagan. philosopher and peasant alike. By the Stoic the bath was deemed essential to virtue, and by the Epicurean to happiness. It is said that St John, the beloved apostle, took the bath at Ephesus, and he lived till he was about a hundred years of age. And of St John Chrysostom, it is related that after he was banished from Constantinople in the fifth century, he lamented the loss of his bath in these terms :—" I am a mere cinder, consumed in the fiercest furnace of fever, without baths or physicians."* Disraeli observes that "Moses and Mohammed made cleanliness religion." And it is a well-known fact that temperance and baths are religious obligations with the Turks. So much so, that the criminal condemned to death in Turkey is as much entitled to his weekly bath as he is to his daily food.

The first and not the least important use that I shall mention of the hot-air bath is Personal Cleanliness. And this effect, I am bold to say, it accomplishes in the most thorough manner, and in a way that no other detergent can do. No one can take a Turkish Bath for the first time but must be astonished at the quantity of unnecessary cuticle which we carry about with us. We have two or three skins too many, and there can be no doubt that the Turks as a nation are far before us in the matter of bodily purification. The cuticle is like a close-fitting tunic, drawn over the whole surface of the body, and a dirty cuticle is like a dirty shirt. Multitudes of people of the upper and middle classes in this

* See Dr Tucker on the Reformed Roman or Oriental Baths.

country fancy that they do not need the bath, because they wash and are clean, and because they have no dirty work to do, like the lower classes. But what would they say if we were to tell them that they are not only quite mistaken, but that they of all people have most need of the bath, and that they are really the dirty classes. Let them not be offended, or suppose that we have become an enemy because we tell them the truth. The moustached mirror of fashion appears fair without, while the hard-working coalheaver appears externally filthy. But, view them in the light of the preceding chapter, and then, physiologically considered, their condition becomes relatively reversed. The working man sweats at his work, and to a great extent gets rid of his uncleanliness, while the gentleman who has nothing to do, eats, drinks, and perspires not as he ought, retains in a great measure within his system the bodily impurities which ought to have been excreted and drained off. Need we wonder that the one looks stout and healthy, while the other appears pale and delicate? When the working man has worn his shirt for a week, its odour is then intolerable; and the reason is, that his skin is an open drain, and the impurities of his skin have been drained off by abundant perspiration. But this could not be the case with many other classes of the community, because their skin is a covered cesspool, an impeded drain; and the foul matters are retained within as the future and fruitful generators of many direful forms of disease. The one class is really clean, the other only looks so.

In a periodical published in America, the following

lively little satire appeared some time ago :—" A man gets up in the morning, washes his hands and face, and pronounces himself clean, and eats his breakfast with great complacency. And clean he would be, were his body composed exclusively of hands and face. Day after day he performs this partial ablution, and conscience never whispers to him that he is misusing the other members of his body most abominably. His head aches, his feet swell, he feels unaccountably uncomfortable, and he never dreams that all this is caused by an obstruction of the pores. He cannot understand that health and good looks depend materially upon general and habitual cleanliness. Cleanliness is not a negative but a positive virtue. What is it that makes the Turks such graceful and handsome men, and the Turkish women so exquisitely lovely? Nothing in the world but their frequent use of the bath. And we verily believe that the truth and honour for which the Turk is proverbially celebrated have more connection with his cleanliness of body than nine persons out of ten would imagine."

Although it has been dinned into our ears from our earliest days that cleanliness is next to godliness, yet it is a fact that real personal purification is not sufficiently attended to. The care of the soul no one denies to be a duty, but the care of the body, though not ranking equal in importance with the other, is a duty too. Is not the body God's gift and God's work? We pay far more attention to cleanliness of our garments, and to the decent appearance of our raiment in the eyes of the world—but " is not the body more than raiment?" Indeed, not only has cleanliness very intimate relations with godliness, but, as Dr Gairdner, in his recent work on "Public Health, in Relation to Air and Water," says, "it is a part of godliness itself." And what was the method by which the very idea of Purity and Holiness, as an Attribute of the Divine Being, was instilled into the minds of the rude and ignorant Israelites after their deliverance from Egyptian slavery? And what were the means by which moral purity, as a duty of man, as well as a perfection of Deity, and a requirement exacted by the True God of all his worshippers and servants, was conveyed into their impure and depraved minds? Was it not, in a very great measure, as has been clearly demonstrated by the author of the "Philosophy of the Plan of Salvation," by means of the numerous ordinances and multiplied ceremonies respecting personal purification and bodily cleanliness, and the connection instituted between them and fitness for engaging in religious services? And although these ordinances have passed away, the care of the body, which is God's gift, and a regard for its purity, ought not to fall into desuetude. But how many neglect it, and appear content, not only as a consequence to drag on a sickly existence of impaired bodily health, but to neglect a plain religious duty? To show how much religion, in its relation to the body and to bodily purity is neglected, I need only instance a case referred to by a country clergyman, who but the other day was spoken to about a man in his village who was supposed to possess great spiritual attainments, when the minister quaintly remarked, that one essential element of spiritual Christianity was wanting—namely, personal cleanliness. Personal ablutions were enjoined upon the Jews by the ceremonial law, and the practice of bathing, and washing, and cleansing the body—a religious duty far too much neglected by Christians—is often enjoined in the Old Testament as a remedy in the cure of certain diseases. The Romans evidently understood the reactive effect of bodily purity upon the salubrity of the mental and spiritual part of our being, as is illustrated in the saying, " *Mens sana in corpore sano.*" Socrates, *apud Platonem*, says, " Bathing renders a man pure both in soul and body." And Thomson, in his " Seasons," says :—

" Even from the body's purity, the mind
Receives a secret sympathetic aid."

Mr Urquhart has an interesting passage on this point, in which he says :—" Where Christianity first appeared, cleanliness, like charity or hospitality, was a condition of life. Christ and the Apostles went through the legal ablutions. When the relaxation took place at the first Council of Jerusalem in favour of the Gentiles, these points could never have been raised or called in question, for in this respect the habits of the nations were in conformity with the Jewish law. Reference is made to it in the fathers, not as a practice only, but as a duty. In the primitive Church of England, the bath was a religious observance : the penitent was in some cases forbidden its use ; but then cold bathing was enjoined. Knighthood was originally a religious institution, and the conferring of it is a church ceremony. The aspirant knight prepared himself by the bath. The second distinction which it is in the power of the Sovereign of England to bestow is ' The Order of the Bath.' Now, the

Sovereign who confers, and the knights who receive the title, never saw a real bath in their lives."*

He gives some curious facts illustrative of the amazement felt by Orientals at the degeneracy of European nations, with all their boasted civilisation, in this respect. He records the exclamation of a woman in Turkey, who refused to let her son go with him to Europe—" Vai ! vai ! are not your ships made fast under my windows, and do I not see how the Franks wash ?" On another occasion, in a steamer, he was defending his countrymen as well as he could, by appealing to the soap, towels, and canteen dishes which were being carried down to the cabins of some English officers, as a proof that his countrymen did not neglect to wash, when one of a group of Turks on board said, "The unfortunates ! they don't know how."

For, let us consider for a moment the question, What is purification ? It does not consist in mere external ablution, or in getting rid only of outward impurities which have been received by contact with external things. No. It is the effectual removal, not of the dirt deposited on the body from without, but of the effete matters, the impurities deposited on the body from within. For it is that which proceedeth from a man that defileth a man. Let us lay it down as an

---

* "What worship is there not in mere washing ! Perhaps one of the most moral things a man, in common cases, has it in his power to do. It remains a religious duty in the East. Even the dull English feel something of this : they have a saying, 'Cleanliness is near of kin to godliness' ; yet never in any country saw I men worse washed, and, in a climate drenched with the softest cloud water, such a scarcity of baths."— SAUERTEIG.

axiom that what proceeds from the body is impure—
that it is a fountain of impurities. And as the skin is
the main organ for drawing off bodily impurities, and as
it is perforated all over with little drains, to the number of seven millions in all, it behoves us to see that it
is in proper working order—to " wash and be clean." It
is the only important organ specially given over to man's
own care. But instead of attending to it, we cover it with
closely-fitting garments, or rather integuments, by which
the excreted matter is never wholly absorbed, but only a
portion of it, and by which, therefore, the perspirable
matter is shut in, and purifying elements are shut out.

That we don't know how to wash ourselves in this
country, moreover, as the Turk said, is really true. For
washing in this country is often the reverse of a cleansing
operation, because such dirt as adheres to the skin, and
might have fallen off from it, as dust, is either hardened
on the skin by cold bathing, or, as Dr Brereton says, is
re-dissolved, and literally washed back into the pores of
the skin. " You may often be conscious of this," he
observes, " by a certain clammy stickiness of the skin
directly after the so-called washing. But just take a
sponge and wash your face over, and see if the water be
not dirty. To proceed to wash yourself, therefore, in
that water is simply absurd. In the Turkish Bath there
is no such operation as this. There is no basin or vessel
of any kind whatever, but a continual stream of fresh
water is passed over the body, which removes all the
filth that has been deposited upon the skin. Let the
fairest skinned person take a bath (or rather four or five,
for the first is not effectual), and he will be astonished

at the black rolls clinging to his person after the shampooing, and this it is which removes from the pores what the previous process had carried to them."*
Enough, I trust, has now been said to establish the claims of the hot-air bath as a purifier—a purifier of the blood, a purifier of the person, a purifier of the whole inward man. This is the bath by which all impurities are brought to the surface, ejected, and effectually removed—the bath which makes a man " clean every whit."

In prescribing rules to Christians in ancient times on the subject of bathing—and bathing in the hot-air bath, be it remembered—Clemens Alexandrinus, in the second century, gives the four following reasons :— Cleanliness, health, warmth, and pleasure. We have begun with the first of these, namely cleanliness ; and now we proceed to speak briefly of the last,—namely pleasure. To consider it simply as a luxurious indulgence, is doubtless placing it and its claims on the lowest grounds. But how few luxuries are there like this, not merely innocent, but highly beneficial. As a positive luxury, it is cheap and easily accessible. The enjoyment to be derived from it is hardly to be realised by the imagination of those who are strangers to it. It produces a harmonious and healthy action of every organ, it completely balances every function of the body, it braces the nerves, it produces a suppleness of the muscles and joints, a firmness of the tissues, an elasticity in the whole frame, tending to a delicious elevation of the spirits and serenity of the mind. Even in this light,

* Lecture on the Action and Uses of the Turkish Bath.

then, the lowest in which it is possible to regard it, namely as a luxury, of how many attractions is it possessed!*

A nation without the bath is deprived of a large portion of the innocent enjoyment within man's reach; and in those countries where it exists, it is not wonderful that the people should become passionately fond of it. An American writer, in giving an account of his sensations after a visit which he paid, in company with a party of friends, to the *Hamâms* at Constantinople, says, "Their skins looked so white, their lips so red, and the delicate rosy tinge of their cheeks so tempting. Even old age seemed to bloom again, while a universal exhilaration took possession of us all. A delicate tissue of softness seemed to envelope my body, and a wonderful feeling of amiability towards all the race of mankind glowed within my breast. Every motion gave me pleasure, and I could scarce recognise myself."

Not only, then, is it a salutary and innocent pleasure, a luxury which we may safely and easily indulge,—it appears also to be a means of promoting personal beauty. Besides being a purifier of the blood, it is also a beautifier of the person. Its natural tendency, as every one at all acquainted with it well knows, is not only to strengthen and invigorate the frame, but also to clear and purify the complexion. Not only does it preserve the health, but it heightens every personal charm. The Irish peasants frequent the rude hot-air baths in the Island of Rathlin before the annual fair of the neigh-

* See the subject further discussed in Culverwell on the Utility of Turkish Baths.

bourhood; and it is particularly remarked that the Irish girls seek that occasion in order to improve their beauty and complexion. John Bunyan seemed to have an excellent idea of the value of hydropathic appliances for this purpose, at the same time that he evidently considered the bath as a means of preserving the body in sound health, and as a most valuable restorative when jaded and worn out with fatigue.

In his "Pilgrim's Progress," we find that when Christiana and her children had through many trials and difficulties reached the house of the Interpreter, they were there comfortably lodged for the night, and saw many instructive pictures that were well fitted to be useful to them during the remainder of their pilgrimage; and when they were about to take their departure in the morning, they were not permitted to set out until they had complied with the rules of the house in reference to the bath. Accordingly they were led out into the garden where the bath was placed. In this pleasant situation it was associated with every thing that was fitted to please and delight the senses, in the varied beauty and fragrance of the flowers and the sparkling verdure of the foliage, all harmoniously blended together, and admirably adapted to infuse a calm and sweet repose in the soul. From the beautiful description he gives of the scene, it is quite evident that Bunyan with all his dreams never dreamt of a bath in a cellar, underground, like many of the baths of the present day, and ventilated, as they must necessarily be, with unwholesome and foul air, tending to do harm rather than good. But—to return to our pilgrims, whom we left in the garden—we are next told that they went all of them into the bath, and when they came out they were not only sweet and clean, but also much enlivened and strengthened in their joints; so that when they returned out of the garden from the bath, they looked fairer a deal than when they went in, and the Interpreter took them and looked upon them, and said unto them, Fair as the moon.

It is evident, I think, that the illustrious dreamer must have had some acquaintance with that renovating process which is the result of hot air in bathing, otherwise he would not have employed language so appropriate and so applicable. His description will readily be recognised as correct, and the results which he mentions will be felt to be matter of every day's experience to those who have the courage occasionally to leave their business for a short time, and promote their own health and comfort by visiting the Turkish Bath.

But the Irish peasant girls and John Bunyan are not my only witnesses on this point. I call other testimony. Dr Brereton, in his lecture, says that, under the use of the Turkish Bath, "the complexion becomes clearer, the eyes brighter, and the person positively fragrant. Homer does not exaggerate when he describes Achilles, on issuing from the bath, as looking taller, fairer, and nearer to the immortal gods." Dr Barter also, in his pamphlet, speaking of the absurd and foolish idea to which some persons came, on not feeling so comfortable as they had anticipated in their first bath, from the perspiration not flowing freely, the skin requiring training to enable it to obey the call to perspire—

that the bath did not suit them, and so, not only depriving themselves of the benefit, but prejudicing the minds of others by the evil report which they industriously circulated far and wide respecting it—observes that he had seen children of two days old and people of ninety years in the bath, and they all enjoyed it, and that none should be discouraged by unpleasant sensations produced at first, as he had always found that it was in such cases that the bath proved ultimately most beneficial. And then he goes on to take the case of a girl fifteen years of age, who was naturally anxious to look as charming and attractive as possible, and considers that she could not adopt a more effectual mode of enhancing her charms than by taking the Turkish Bath. To the young it was most beneficial in assisting the growth and development of the body, a high degree of temperature being the great stimulus to growth in the animal as well as in the vegetable world. Wherever the bath was a national institution, the hair of the women was peculiarly luxuriant and beautiful. And he concludes by saying that he can vouch for it, that the use of the bath purifies the skin, and renders the complexion more delicate and brilliant. Now, while bearing my full testimony to the truth stated by Dr Barter, I think it would have been better and more appropriate if he had gone higher up the scale, say to four times fifteen, or even higher, and employed his eloquence in showing, as he might easily have done, the powerful influence of this bath in refreshing, renovating, invigorating, and restoring the fading beauty and energy of her who had borne the burden and heat of the day, and for sixty or seventy years " braved the battle and the breeze of life's stormy seas." The Doctor must have forgot himself altogether when adducing this illustration. His girl of fifteen is but a tiny skiff, just spreading her canvass in a smooth sea. The other is a stately frigate, that has buffeted many a storm, and nobly breasted many an adverse wave, and therefore deserving of double honour. It is reasonable and proper then that this latter should be invited to avail herself of the privilege of this powerful renovating process, which makes the old young again, and enables them to renew their youth, in order that the precious and valuable life may happily be prolonged to a good old age.

There might have been observed in the public prints recently a notice of a young woman who had been brought before the Insolvent Debtor's Court, who gave herself out to be, by trade, an enameller of ladies' faces and a dealer in cosmetics, and stated that she was in the habit of receiving from five to twenty guineas, and sometimes more, for enamelling a lady's face. Now, how foolish and absurd is the practice of those persons who vainly try to enhance their charms and their beauty, by having recourse to pernicious and poisonous substances, such as paints, cosmetics, hot-irons, and enamels. These appliances must effectually destroy what little beauty or comeliness of countenance they may have previously possessed. In consequence of the stoppage and destruction of the pores in that part of the skin to which they are applied, they very soon produce in the human face divine unnatural wrinkles and premature decay, and cause it even in youth to assume the aspect of advanced age. If ladies would study economy, and their own happiness and comfort, and at the same time preserve and prolong their beauty, they may accomplish their desire, and have their faces enamelled in a natural way, and at a mere tithe of the expense, by visiting the Turkish Bath. Darwin, in his " Zoonomia," says, " To those who are past the meridian of life, and have dry skins, and begin to be emaciated, the warm bath, for half-an-hour twice every week, I believe to be eminently serviceable in retarding the advances of old age." If this be true of the simple warm bath, how much more true is it of that best of all warm baths, the hot-air Roman Bath! " In the bath," says Dr Wilson, " there are no wrinkles, and no decrepit age ; the skin becomes firm and elastic ; it recovers its colour and its smoothness ; it fits close to the muscular frame beneath ; its hues are selected from the palette of youth ; as the skin regains its health, the hair returns upon the scalp of the bald ; and white hairs, which have crept untimely and unbidden among the locks, shrink away."

Thus I have shown the virtues of the bath in regard to cleanliness and to comeliness—that it is a Purifier of the body and a Beautifier of the person. I now go on to speak of it as a Preserver of the health, and a Fortifier of the frame against cold. Health and Warmth are among the purposes for which, it will be remembered, old Clemens Alexandrinus recommends it. Into the groundless fear of catching cold after it I shall not now enter, but reserve the answer to this and all other objections to a subsequent chapter. How, let me ask, are animal heat and bodily health maintained? There is but one proper, praiseworthy, and permanent method of maintaining bodily heat, and that is, not by stimulants, nor by fires, nor by heaps of clothes, but by the combustion of carbon, from its union with the oxygen of the atmosphere, which takes place not only in the lungs, but all over the surface of the body. A person with a dirty skin may clothe himself with loads of garments, and yet he will be found to shiver beneath them all, for there is in his case no free passage for the pure oxygen that is in the air to enter the seven millions of closed pores in his skin, or for the carboniferous matters contained in the tissue of his skin around the pores to escape, and come in contact with the oxygen in the atmosphere. Contact and combustion between the two are effectually precluded, and, in a great measure, prevented, so that *Heat* is not generated, and *Health* is impaired. No wonder, therefore, that the person shivers. What, then, is necessary to prevent coldness? I answer, a more active circulation of the blood in the extremities, and on the whole surface of the body. How, then, can the circulation be rendered less languid? I answer again, by admitting more air. Remove the dead scales from the skin, and pump out and extract from the pores the effete matter which blocks them up, which keeps carbon in and oxygen out—and this, I unhesitatingly affirm, you can secure most effectually by means of the hot-air bath. Can the circulation be effected *without* air? No. If you cut off the atmosphere from the lungs, all the power of the heart will not be able to drive a drop of blood through that organ. And so it is with the skin. It is an Air-Breather; and just as the lung, with its millions of cells throws off carbon and takes in oxygen, and is thus the great laboratory of heat and life for the interior of the body, so the skin over its whole surface breathes, exhales carbon, and admits oxygen, less thoroughly and directly than the lungs, but still to such an extent that, when kept properly clean, and in good working order, it generates its own heat abundantly. For it is not the clothes that create heat. They have no more heat in themselves than a stone has. But they retain the heat which the body has generated. Hippocrates, when dying, consoled those who stood by and lamented his decease, by saying, that he left three excellent physicians behind him,—Air, Water, and Diet.

The effect of the bath, then, is to generate heat, develope warmth, preserve the health, and promote the comfort of the whole frame. It produces an impatience of clothing, so that after a few baths you are glad to throw off some of the burden of clothes which you may have been formerly in the habit of wearing, but which you were afraid of diminishing. Now you gladly rid yourself of a portion of the weight, and you delight to have the oxygen freely circulating near your person. The valetudinarian can now walk forth without fear and danger of cold at every passing breath. This comfortable feeling of security against cold may not be felt immediately, but it will certainly be felt after a few baths. Nay more, cold will not be felt in the same manner, or to the same degree as before. It may be as cold as it used to be, but you do not feel that it is so. You rather feel a pleasant glow; you rejoice in being abroad in the open air, and you feel that you are inhaling the air at every pore. Instead of being benumbed and shivering, you now positively delight in exercise in the bracing cold. At the same time, of course, attention to the clothing must not be neglected. " Every one whom I meet to-day, cloaked up to his nose," says Dr Brereton, "tells me that it is a cold day; I am not sensible of it, though I was formerly as sensitive to cold as most people, and once was even considered as consumptive. I assure you, that so long as I take my bath once or twice a week, no degree of cold affects me, let me clothe as thinly as I may."

There are various other excellent uses and beneficial purposes to which the bath is applicable. It is a cathartic, a diuretic, a detersive, and a tonic. It is a preservative, a preventive, and a curative, being in its relation to disease, and in the hands of a physician, a powerful prophylactic and therapeutic agent. But this is too important a part of the subject to be entered into or discussed at the end of a chapter. I will therefore reserve it for a separate chapter by itself. Meanwhile, it may be noticed here, in conclusion, that the bath is strictly promotive of temperance. Mr Urquhart, in his second volume,* says, "I know of no country in ancient or modern times, where habits of drunkenness have co-existed with the bath. Misery and cold drive men to the gin-shop; if they had the bath—not the washing-tub, but the sociable hamâm,—to repair to, this, the great cause of drunkenness, would be removed; and if this habit of cleanliness were general, restraint would

* Pillars of Hercules, vol. ii. p. 78.

be imposed on such habits by the feelings of self-respect engendered." And, again, he says, " Do not run away with the idea that it is Islamism that prevents the use of spirituous liquors ; it is the bath. It satisfies the cravings which lead to those indulgences, it fills the period of necessary relaxation, and it produces, with cleanliness, habits of self-respect, which are incompatible with intoxication : it keeps the families united, which prevents the squandering of money for such excesses. In Greece and Rome, in their worst times, there was neither blue ruin nor double stout."

If such be a correct view of the case, as I believe it is, then the duty of every true friend of humanity is obvious. A war has been waged against drunkenness for many years. Holidays and recreations have been provided for the people. Abstainers have laboured in this field through evil and through good report, in order to rescue their fellow-men from the fearfully degrading vice of intemperance, with its long train of untold evils and miseries. Societies have been everywhere organised, large sums of money have been expended on prize essays, lectures, &c. In a word, everything has been done that could be done in the circumstances—except one thing, the erection of baths for the people. The evil we see to be nearly as rampant as ever. The large amount of good which has undoubtedly resulted from the extensive machinery which has been in active operation for so many years, is but as a drop in the bucket compared with the extent and magnitude of this monster evil—the bane and curse of our country.

Seeing, then, that the means are not adequate to, nor

commensurate with the necessities of the case, could not a part of the sums annually expended for the accomplishment of this great end be directed into this new channel which has just been opened up to philanthropic exertion—I mean in the erection of Turkish Baths for the working classes and for the poor? Were some benevolent person to bequeath a sum of money for the purpose of erecting a Hamâm for the people of Edinburgh, open to all, and free of charge, as in ancient Rome, or at a merely nominal charge, as in modern Turkey,—every Saturday night, or every night of the year, what a boon to the people it would be! And they would very soon come to value it, to find the benefit of it, and to avail themselves of it as a blessing to them.

I perceive from the public papers that both the English and the Irish are far ahead of us in this matter. In Bradford, and in various towns in England, the working-classes have taken the matter into their own hands, and have founded and maintained joint-stock bathing companies, by means of which an artisan can obtain an excellent hot-air bath for the sum of threepence; and as regards Ireland, a subscription has been opened in Cork for the purpose of providing Turkish Baths for the poor. This is as it should be, and why should the people of Scotland be behind in this great and noble enterprise of providing these temples of health and temples of temperance for this class of the community? Thus may we hope that the victims of intemperance may be induced to forsake those accursed whisky-palaces, which are the sources of destruction and death.

The Turkish Bath is social and sanitary in its operation. It utterly destroys the craving for strong drink ; it is an effectual weapon with which to fight against this monster evil. It is a sure and certain and easy cure against an appalling and enslaving habit.

The illustrious Bacon, it is said, lamented the loss of this bath to European nations. It has been truly a great loss. But, now that it is being revived and restored, what a precious boon it may prove! If sultans and heathen princes built and endowed such places for the honour of their name, is there not enough of Christian philanthropy to be found among ourselves, to prompt us, with the true spirit of benevolence to our fellow creatures, to go and do likewise? There can be no doubt that if the object desiderated were fully known and understood among the masses of the community, and particularly among the monied classes, the end might be speedily accomplished, not upon the low ground merely of worldly honour, or for worldly gain, but for the glory of God and the good of our common humanity. They would have the satisfaction of knowing that they were lending a helping hand in forwarding a movement which would effect a great moral revolution in our social and domestic habits, for which unborn ages would arise and call them blessed.

I have said the Turkish Bath completely overcomes and destroys the craving for strong drinks, and I am able to furnish evidence of the assertion. Individuals have come to the bath for the avowed purpose of getting cured of that, which had become in their case a disease, so confirmed did it appear to be, and have experienced the most beneficial effects. Before I introduce their testimony, which will be in their own words—before giving their letters, I would notice the fact which was lately brought under my attention, of several horse jockeys at Sydney, who trained themselves in the bath, and who all bore testimony to the fact, that though they dropt weight much faster than by the usual process of gin-drinking and blanket-sweating, they became stronger instead of weaker by bathing, one of them having bathed four times a-day—but, it was added, his appetite increased, he had no desire for drink, and he was firmer in flesh than before.

The first witness to the beneficial use of the bath in respect to temperance and sobriety, says—" Dear Sir,— Having used the Turkish Bath pretty steadily for the last three months, I beg to notice one effect I have experienced from its use. I may mention that before taking the bath, it was a quite common thing for me to take a glass of whisky and water, or toddy, at night before going to bed; but since using the bath I have not tasted whisky, or any kind of spirits—and that not from any moral effort on my part, but, simply that I have not felt the slightest inclination for anything of the kind. I think the bath has produced a feeling of such comfort, a lightness both mental and physical, as to render a stimulant of any kind quite unnecessary— hence the desire for it vanishes. Such has been my experience."

The second is as follows :—" Rev. Sir,—I have much pleasure in giving you my best thanks for putting it in my power to avail myself of a trial of the Turkish Bath.

At the suggestion of Dr Lawrie, who recommended a course of 24 baths, as a cure for what I call chronic drunkenness, I have now had 19 baths, at three per week, with very decided beneficial effects. I have during that time had four attacks of the—in former times—uncontrollable, and always overpowering, craving for strong drink; and I have been enabled by God's grace, and His blessing on the Turkish baths, to resist the temptation to indulge. I account for the return of the craving, at so short intervals, from the fact that it has not been gratified. I may state that for a long time I have hated strong drink, for it was ruining me, body and soul, robbing me and my family of every thing worth living for in time, and left nothing in prospect but an undone eternity; and yet, so strong was the craving when it came on, which was every three or four weeks, that I could not resist. I repeat it, and I do it advisedly—I could not resist. But I wish not to be misunderstood. I am not cured, but the craving is not so strong, does not last so long, and I believe will soon be so slight as to be little felt. On the whole, my general health is better. I sleep, eat, and work better than before attendance on the baths. I believe, however, it will only be by constant watchfulness and prayerfulness, and keeping far from the very appearance of evil, and by the grace of God, that the victory will be complete. Sir, if you think this will be of any service, use it as you choose. And I will be happy if any information and advice I can give will help to save any poor wretch from this awful vice.—I am, sir, your grateful and humble servant."

CHAPTER VII.

PROPHYLACTIC AND THERAPEUTIC BENEFITS.

I propose in this chapter to show the value of the Turkish Bath, as an agent at once prophylactic—that is, preventive of disease; and therapeutic—that is, curative of disease.

Generally, it may be observed that prevention is better than cure. Now, that the bath is a preventive of disease, and an admirable sanitary agent, is clearly evident from the fact that it effectually produces and maintains a healthy condition of the skin. The importance of a healthy skin, as conducing to the health of the entire organism, has been already demonstrated. The skin is worth preserving in a healthy state; and the bath not only aims at this result, but accomplishes it. It prevents disease by fortifying and hardening the bodily constitution against miasmatic influences and zymotic affections, against variations and vicissitudes of temperature, against diseases of the nutritive and digestive organs, and against diseases of the cutaneous system, the muscular system, the nervous system, and the reproductive system.

As an eliminator of noxious substances from the blood and from the bodily tissues, by drawing peccant humours and morbid matters to the surface, and ejecting them through the pores, the hot-air bath is invaluable. Doubtless this is the reason why the general health of those communities where it is frequently used stands so high. Thus we are informed that many diseases which are common among us are scarcely to be met with in Constantinople; that gout, gravel, and stone, are medical curiosities; that rheumatism, *colica pictonum*, or lead-colic, and chronic skin diseases, are very uncommon; and that, notwithstanding the want of sewerage, drainage, or good nourishment, the population appears to be healthy.* Again, Dr Millingen says, "As to the application of the bath in the prevention and cure of diseases, the working classes among the Turks know of no other means of prevention, on feeling indisposed, but the bath. In the numerous cases arising from sudden changes in the temperature of the body, a copious perspiration, which a stay of more or less duration in the Calidarium is sure to occasion, does, in the great majority of cases, restore the body to the equilibrium of health. After over-exertion, again, the bath is had recourse to. In short, it is looked upon so much in the light of a panacea by the lower orders, that they hardly ever dream of consulting a physician when taken unwell. If the bath fail to cure them, nothing else will. This prevailing conviction accounts, in a great measure, for the total absence of dispensaries and civil hospitals, not only in that large city, but throughout the empire."

The history of disease in ancient times confirms the

* See Haughton's "Facts and Fallacies on the Turkish Bath Question."

conclusion that this bath is powerfully prophylactic and preservative of the public health. Small-pox, a skin disease which was for ages the scourge of Europe, appears to have come in just as the Thermæ of the ancient Romans went out of general use. That disease, together with scarlatina, measles, and other diseases of the zymotic type, as also cholera—dire scourges now—were altogether unknown to the ancient Greeks and Romans. So that it would almost appear as if these diseases were sent as a penal infliction for the criminal neglect of personal purification, so much more common in these later than it used to be in the earlier ages of the world. When the population of Rome was at its height, there were between 800 and 900 baths, but we do not read of one hospital. It is even said, that for several hundred years there was no physician there. Baths and gymnasia were their medical and sanitary institutions. And if we extend our survey to those places where one part of the population habitually use, and another part habitually neglect this bath, we shall arrive at most striking results. In Cyprus, for instance, the population is partly Mohammedan, and partly Christian. The former take the bath as a religious rite, and are singularly exempt from pulmonary consumption; the latter neglect the physical duty of personal purification, and are as subject to this dire disease as other European nations. Mr Urquhart, whose observation has been very extensive, says,—" Where the bath is the practice of the people, there are no diseases of the skin. All cases of inflammation, local and general, are subdued. Gout, rheumatism, sciatica or stone, cannot exist where it is consecutively or sedulously employed as a curative means. As to consumption,—that scourge of England—that pallid spectre, which sits by every tenth hearth, among the higher orders,—it is not only unknown where the bath is practised, but is curable by its means."

It is a lamentable fact, that in this country a very large proportion of human beings die before they reach the age of five years; and this melancholy circumstance cannot be too strongly impressed upon the heads of families, and especially upon mothers and nurses, and also upon the learned gentlemen who are the professed guardians of the public health. Experience has abundantly proved, that many of those formidable diseases which commit such havoc upon the tender lives of infants, might be more easily managed, if not entirely prevented, were the attention of medical men and others directed more to skin treatment than dosing with poisonous and nauseous drugs, and other cruel appliances. Were this done, doubtless many of those innocent victims would be preserved from much unnecessary suffering, and also from a premature grave. Professor Simpson remarks in a recent work, on this deeply interesting and important subject, that the attention of medical men ought to be constantly fixed on this point, with a view to discover some prophylactic means by which such fearful and fatal maladies might be arrested, and rendered comparatively mild and safe. The learned Professor very justly observes, that if, by vaccination during infancy, medicine has devised prophylactic means to arrest the ravages of small-pox, may it not yet devise some analogous means also to arrest the ravages of scarlet fever and measles, of hooping-cough and typhus fever, and perhaps of the whole class of non-recurrent diseases, by artificially producing these several diseases in a mild and safe form, by inoculation and imitative medication or otherwise? "Let us, at least," he says, "not sit indolently down and argue ourselves into the belief that it is impossible to attain such results. The conquest of small-pox seemed to our forefathers a hundred years ago as impossible as the conquest of these maladies can look to any one now; and yet we all know that the subjugation of small-pox was effected by the genius of one man, and the devotion of one mind to its accomplishment." He farther declares, that he entertains an ardent belief that medical science may yet devise measures, prophylactic, perhaps, rather than curative, to stay the great destruction of human life prevailing amongst us from consumption and other diseases. Surely every one will say, after reading these ardent and benevolent aspirations of the learned—and, I may say, "the beloved physician," that what he looks forward to and hopes for is a blessed consummation most devoutly to be desired.*

Now, a question naturally suggests itself here,—Has this specific remedy been yet discovered or generally adopted? Let the bills of mortality answer. Let the innumerable families which have been visited and desolated by the ravages of that fatal malady, scarlet fever, alone, answer the question. It is very true, that a variety of remedies, said to be specific, have been from time to time recommended and introduced with a high-

* ee Simpson's Obstetric Works, vol. ii. p. 483—On the Possible Prophylaxis of Scarlatina and other Diseases.

sounding flourish of trumpets, but which, when brought to the test, have proved miserable failures, with the exception of one indeed, namely, belladonna, which, according to some authorities, is a powerful remedy against scarlet fever; and that it possesses a very wonderful power in certain cases of that disease, there can be no doubt. I speak from my own experience of it for many years,* and I could easily bring forward ample

* That it is an invaluable and powerful curative agent, the following cases will show. I was in attendance on a married female in 1840, who had been suffering from acute inflammation of the throat, of a very obstinate character. All the usual allopathic remedies recommended in such cases failed to give the slightest relief. Leeches, gargles, blisters, purgatives, &c., were all in due course administered. Another physician was called in without my knowledge; and on again visiting, I was glad to find the patient much relieved. She informed me that she had been advised by a friend, especially as she had not been benefited by the means hitherto used, to call in the physician alluded to, without intending any offence to me; and that he had given her some medicine, one teaspoonful of which effected immediate relief. On examining the prescription I found it to consist of about a sixteenth of a grain of belladonna, dissolved in water, a teaspoonful for a dose, three times a-day. The next day the patient was quite well, and required no farther attendance. I was no less delighted than surprised at such a speedy, effectual, and, as it proved, permanent cure, effected by a minute dose of a medicine which I had never before heard of as applicable to such a case; but, in the course of my investigations, I found that belladonna was a most important homœopathic remedy, invaluable in many diseases; and I afterwards had occasion to know that the physician who prescribed it in the above case, although an extreme allopathist, had derived his knowledge of its efficacy in such cases from a homœopathic source. And this was not the only remedy of a similar kind he was in the habit of using, a supply of which he regularly obtained from Germany.—Scarlet fever prevailed to a considerable extent in Edinburgh at this time, and was very fatal in many families. I remember well a very severe case in a girl ten years of age who was under my care. It seemed as if all the virulence of the disease was concentrated in the throat and the glands of the neck. Alarming head symptoms supervened; the tonsils on

testimony from many eminent medical men on this point. But I merely name it in passing. What I have here to state, then, is this, that a more powerful remedy than any that has as yet been discovered, is to be found both sides were covered with ulcerations ; the rash on the skin was of a dark red ; the fever ran very high ; the pulse about 120, rapid and small. I had used every allopathic remedy I could think of, and had repeatedly applied the caustic to the ulceration in the throat, but nothing seemed to be of any avail. I called on the evening of the seventh day, about nine o'clock—as I then thought for the last time, for I had given up all hopes of the girl's recovery : she was very restless, rolling the head from side to side on the pillow, continually tossing the arms to and fro ; the countenance expressive of great suffering and anxiety ; with delirium. On examining the throat, which I had great difficulty in accomplishing, owing to the tenderness and irritability of the entire mouth, I found it in a state of acute erysipelatous inflammation, parched and glossy, resembling very much that in which the belladonna had been so effectual. I therefore resolved to try belladonna in this case, and ordered a small teaspoonful to be given as soon as the medicine could be procured, and repeated at two A.M. I then took my leave, fully expecting that death would close the scene before morning. On calling next day, however, I was delighted beyond measure, and not more so than surprised, to hear the mother of the child, with a cheerful and animated countenance, exclaim in her own simple and expressive manner. " Oh, sir, there was surely a charm in that medicine you ordered last night the poor child had no sooner taken the first dose than she was relieved, and fell into a profound and refreshing sleep." Sure enough, I found the girl in a very different state from that in which I had left her on the previous evening. The pulse was soft and regular ; the countenance pleased and natural ; the inflammation in the throat almost entirely gone, and the ulcerated tonsils presented a healthy appearance. They rapidly healed, and in a few days the girl was quite well ; and she had no relapse whatever. I prescribed the same remedy in many other cases with marked benefit, and recommended it to several of my professional brethren, urging them to give it a trial. The late Alexander Miller, surgeon, F.R.C.S.E., was prevailed upon to use it in a similar case under his care at the time. On asking what he thought of the remedy, his reply was most emphatic " I have no hesitation in saying that it saved the child's life."

in the Turkish Bath. And of the truth of this, I shall adduce abundant evidence from those who have personally witnessed its efficacy in such cases as those of which I am now speaking.

But it is a remarkable fact, and, especially in reference to medical men, as true as it is remarkable, that the very thing so much desiderated and longed for, if it does not present itself in a fashionable court dress, or if it does not travel along the groove of what is called orthodox or legitimate medicine, is at once denounced as quackery. Whatever runs counter to their preconceived ideas and notions of things is set down as heterodox, and medical men, like other mortals, are often unwilling to enter on the investigation of any new truth solely on its merits, or to subject it to the test of actual experiment in order to ascertain the truth or falsehood of the novelty. Thus has it been with the truth contained in hydropathy,* homœopathy, and last, but not least, with the Roman or Turkish Baths. There are, however, many noble exceptions to be found both in Edinburgh and London, and also in other places, gentlemen of high standing in their profession, who are fully convinced of the great value of this most powerful prophylactic and curative agent, and who have acquired a correct know-

* In his " Horæ Subsecivæ," Dr John Brown, amid some *asperrima verba* which I do not feel called upon to quote, says, " If all that is good in the Water-Cure, and in Rubbing, and in Homœopathy, were winnowed from the useless, what an important and permanent addition would be made to our operative knowledge,—to our power as healers; and here it is, where I cannot help thinking that we have, as a profession, gone astray in our indiscriminate abuse of all these new practices: they indicate some want in us. There is in them all something good."

ledge of its physiological action upon the human frame, having tested the merits of the bath both in their own persons and also upon their patients. And, in proof of this, I may state that several physicians of the very highest eminence in their profession in this city, who have risen above the prejudices of their brethren, not only patronise the Turkish Bath themselves, but send their patients to me for relief in very obstinate cases of disease.

Some, who have given it a fair trial, have favoured us with the result of their inquiries, such as Dr Goulden of London; Dr Erasmus Wilson, whose book on the subject is well known; Dr Barter of Cork, and many other eminent physicians, some of whom I have already had occasion to mention. Dr Barter, in a recent lecture, gives it as his opinion, that the monster evil, by which one-sixth of the human family are carried off, is scrofula, in one form or another. This is one of the most malignant and inveterate of blood-poisons, and wherever there is the least taint of it in the system it is liable to be transmitted from one generation to another. And so far as I have been able to observe, during thirty-five years medical practice, I have invariably found that wherever this scrofulous diathesis existed, whether in old or young, the disease with which they were afflicted was always more unmanageable, malignant, and fatal, than in others where the taint was not present. The following facts are brought out in his lecture. He remarks that scrofula does not exist in the torrid zone, because the skin is kept in a high state of excitement by the temperature, and the light, loose dress worn in hot countries allowing the pure air to have constant access to the pores of the skin. Neither is it found in the frigid zone. This may be accounted for to some extent by the fact, that there is a larger amount of oxygen in the air than exists in temperate climates. But the principal cause is this: the natives of the frigid zone, according to Kane's "Arctic Travels," sleep in huts heated from 91 to 100 degrees—in fact, take a Turkish Bath all night, and in the morning are subjected to the invigorating influence of the intensely cold air.

I might go on at considerable length on this part of the subject, but enough has been said to establish the great practical truth, that in proportion as we attend to the functions of the skin will the health and vigour of the body be maintained, and it will be better fitted to resist the many morbid poisons with which we are surrounded, particularly in large towns, and also to throw off those generated in the system itself. A learned medical writer,* in one of the periodicals of the day, observes, in reference to this point, that contagion in such a state of the constitution as that which the Turkish Bath is fitted to produce, finds no material to ferment upon, so as to assimilate to its own poisonous nature, and by which to propagate its destructive virus. He further observes, that various prophylactics against pestilential diseases have been proposed, but if this conjecture be well founded, there is none equal to the Oriental or Turkish Baths. It is even contended by some that as a preservative it is superior to vaccination, as a safeguard against small-pox, as well as other diseases of the

* See Dublin University Magazine, 1861.

zymotic class, such as measles, hooping-cough, scarlet fever, &c., from which the ancients, who universally employed the hot-air bath, were happily exempt.

Having thus shown that the bath is an excellent prophylactic,—preservative of health, and preventive against disease,—I now proceed to speak of it as a most valuable therapeutic, or curative and remedial agent, in the case of many obstinate and painful diseases. I do not mean to assert that it will cure every disease; that would be absurd. We all know that there are many diseases which no human remedy can reach. Nevertheless, some which have been considered as belonging to this class have been immensely benefited by the use of the hot-air bath. And I have no hesitation in saying that the lives of many such have been lengthened out to a longer period than, humanly speaking, they would have attained, had their systems not been delivered from the morbid and poisonous matters with which they were previously so much oppressed. The bath was to them what the life-boat is to the shipwrecked mariner. It saved them from impending death, and bore them safely to the shore. For there is no doubt that if man were to use the means which God has placed within his reach, his life might be prolonged much beyond what it actually is. We have a clear proof of this in the average extension of life which has taken place since the attention of men has been directed to the conditions that determine health, such as personal cleanliness, proper ventilation, pure air, better food, &c.

In the sequel of this chapter I propose to present to the reader a few facts, drawn from my own experience and that of others, illustrative of the beneficial effects and curative virtues of the hot-air bath. Dr Goulden, of London, says, " I watched the effect of the bath in Bell Street. I went into the bath at such times as that I could observe its effects upon the lower classes, who resorted there in great numbers, not as a luxury, but as a remedy, as they supposed, for disease ; and I consider, however much any one may sneer at my occupation, I could not be better engaged than amongst these people, and studying so interesting a subject, even at some inconvenience. There were often ten people in the hot room at one time, all invalids, and I found them quite willing to tell me all their complaints, and to let me examine them. They were principally artisans, small shopkeepers, policemen, admitted at a small fee. I saw there cases of fever, scarlatina, phthisis, gout, rheumatism (acute and chronic) sciatica, bronchitis, forms of skin disease, diseased liver, dyspepsia, ague, dropsy, with diseased heart, and diseased kidneys." The Doctor is very modest and careful in his statement as to the effect of the bath upon the cases which he enumerates. He observes. " To expect a cure, or even benefit, in all these cases would be unreasonable." Mark his words, " But I found relief produced to a far greater extent than I was prepared for. The most marked relief was found in gout, rheumatism, and sciatica." The Doctor's previous caution and carefulness in what he says renders what follows all the more weighty and worthy of attention. " Bronchitis was relieved at once in many cases ; but it required several baths to cure those cases which were of any long standing. I saw a number of children suffering from scarlet fever, who were brought down daily; the cases, many of which were very severe, all recovered rapidly, and left no serious results. One case of typhoid fever was much relieved. This is what one might expect, because this form of fever is unknown in the tropics. But what struck me as most remarkable was the influence of the hot room in quieting the circulation in some cases of palpitation of the heart." He refers to the soothing effect it had in such cases. Mr Urquhart notices, among the other qualities of the bath, its narcotic effect, and I have noticed patients strongly disposed to sleep in the hot room. Dr Goulden enumerates many other cases that were benefited by these Turkish Baths for the common people in London.

Again, in a description of the bath at Cork, by one who had visited that establishment, he says, " I see patients arrive with haggard looks, sunken eye, bent shoulder, and trailing step. A short time elapses, and I observe the same persons with a bright eye, a clear visage, and an upright gait." Dr Barter states, that " from the earliest period of medical history, the value of perspiration has been admitted as a means of cure in acute and chronic disease, seeing that the unaided efforts of nature usually resort to it : and our medical books abound with compounds called sudorifics and diaphoretics. Just contrast this, that all the efforts of nature are turned from their true conservative course, and are expended in getting rid of the drug, through some other channel, when the skin is in a morbid condition, with the beautiful and simple means of the hot-air bath to accomplish this great object." And again, Dr

Hufeland says, "The more active and open the skin is, the more secure will the people be against obstructions and diseases of the lungs, intestines, and lower stomach; and the less tendency will they have to gastric (bilious) fevers, hypochondriasis, gout, asthma, catarrh, and varicose veins."

But the most remarkable testimony which I have met with is that of Sir John Fife, consulting surgeon of the Newcastle Infirmary, into which he had introduced the bath for the benefit of the patients. "My anticipations," says Sir John, "as to the success of the Turkish Bath in the treatment of disease have been fully realised. The temperature ranges from 130° to 160°, according to the nature of the disease, state of the circulation, and condition of the patient submitted. The extreme heat exerts less influence on the heart and circulation than the ordinary warm bath. In some cases of heart disease, patients have undergone the process with unlooked-for benefit. In dropsy, from liver and kidney disease, the profuse perspirations afforded more relief than could have been attained by medicine in the same time. In rheumatism and skin diseases it is invaluable. Internal congestions, chronic and scrofulous inflammations, are relieved, and an equality of circulation brought about by it. I have been restored to youth by being roasted alive in this bath of the Romans."

After these general testimonies in favour of its curative powers, I now proceed to adduce some particular cases which have come under my own observation. And first, I shall mention the case of a lady who had been suffering from acute rheumatism from August 1860 to September 1861. Every remedy that medical skill could devise had been tried, and also the water-cure treatment at a hydropathic establishment, without any satisfactory result. She presented the following appearance :—the countenance anxious, and expressive of severe suffering, also unnaturally red, with a bright hectic flush on both cheeks; the tongue very red and dry, indicating extensive mucous irritation and complete derangement of the digestive organs; no appetite; and the whole complicated with a distressing leucorrhœa of a very acrid nature, which must have existed for some time prior to the severe attack of rheumatism in 1860, which had evidently weakened and reduced the system very much, and in all probability was the remote and exciting cause of the whole train of symptoms which supervened. The disease affected the large joints of all the extremities, and also both hands and fingers. The whole were very much contracted, swollen, and inflamed, and very painful even to the touch. The veins, particularly of one leg, were enlarged, hard, inflamed, and painful. The body could not be moved in any direction without causing excruciating suffering. The patient was entirely helpless, and could not even turn herself in bed. And some of the former medical attendants had expresed an opinion that the lower part of the spine was seriously affected, and that consequently no remedy would be of any avail. Thus the treatment was commenced under very discouraging circumstances, and with doubtful prospects of ultimate success, being a truly formidable case. Without entering further into detail, I will now allow the lady to speak for herself:—" *Bridge of Allan, September* 11, 1862.—My illness commenced in August 1860, with complete prostration of strength, and a swelled or varicose vein became much inflamed and painful. I was confined to bed with it for three weeks, during which time my arms became stiff and painful. In September I went to Rothesay, hoping change of air would restore my strength; but the journey was too much for me in my weakened state, and I became very ill with fresh attacks of pain and fever. After my return home in December, the winter being uncommonly severe, I lost the power of walking from inflammation of the knee-joints, and I was also unable to take my food without help, continuing in that state all the summer. In September 1861, I was induced to try Dr Lawrie's Turkish and Medicated Baths, which seemed at first to make me worse rather than better. I was almost in despair, and envied every poor beggar that had the use of his limbs. About March I began to improve a good deal; and when I returned home in April, people were all astonished with my rosy healthy appearance and improved powers of motion. Still I moved about with pain and difficulty, but by perseverance with the baths, I am now, I hope, fast progressing towards perfect recovery. A more apparently hopeless case than mine can hardly be conceived. Dr and Mrs Lawrie kindly encouraged me to persevere with the baths, otherwise I believe I should have been confined to a chair for life. Thanks to my heavenly Father for blessing the means. I am, dear sir, yours most gratefully." Three months afterwards I received a letter respecting her, of which the following is an extract :—" Mrs —— continues to improve. You would be quite surprised to see her go up and down stairs alone, and so quickly too."

This was certainly one of the most desperate, as it was one of the most obstinate cases that ever came under my observation, and the result is therefore all the more striking as a triumphant proof of the success of the Turkish Bath in medical treatment. The patient was at first so helpless that she had to be carried from the carriage into the baths. The treatment was energetically and perseveringly carried out for seven months, with generally two Turkish baths and four medicated baths per week, which seemed at the first rather to aggravate all the symptoms. We were all at our wits' end, except the poor sufferer herself, who with many tears continued to hope against hope, and thus the treatment went on for about six weeks until slight symptoms of amendment were developed. The appetite began to improve, the joints were a little relieved and less painful, the countenance assumed a more healthful appearance, and became animated with the speedy prospect of returning health. The fingers could now hold a pen, and notes and letters were written to a large circle of friends, and she could occasionally play a tune on the piano. She could now stand on her feet without help, and soon began to walk through the house with a crutch, and all the other symptoms with which the case was complicated began to disappear. In fact her person was so greatly improved and invigorated that those who had seen her previous to the treatment were surprised with the wonderful change which had, by the blessing of God on the baths, been effected.

I shall add only one other case:—"CARDOW, *July* 1862. Having had a severe attack of articular rheumatism, by which I was in a helpless state, confined to my bed and sofa for three months, I was advised by my medical attendant to try the Turkish Baths; and so soon as I could stand the journey of 200 miles, I came under the care of Dr Lawrie, and, after undergoing a course of fourteen baths by his special direction, I rejoice to say that I have benefited to my entire satisfaction, and now return home well, and hale and hearty. The attendance and comforts in this establishment have only to be tried to be appreciated by any reasonable party." Many such cases might be given, which, however, would only swell the book to too great a size, and I think it best to content myself with giving the above case in the writer's own words, without note or comment. I shall now proceed to add a few testimonies from eminent medical authorities, to the beneficial results of the hot-air bath in affording relief and cure to this class of diseases.

Dr Brereton says—"In no disease are the effects more magical than in rheumatism, often when it has defied every other treatment; since my last lecture, several cases of rheumatism, and of years' standing, have been cured in this town (Sheffield). I have just now recalled to my memory a terrible case of acute rheumatism, or rheumatic fever; all treatment had failed to give any permanent relief; the patient with great difficulty could be carried to the bath; he roared with pain if even a finger were moved; you will scarcely believe the result: he WALKED home after the first bath. All forms of neuralgia, which arise from irritation of the nerves by superfluous or effete matter in the blood, such as rheumatic sciatica, many cases of tic doloureux, &c., the bath will certainly cure. The high temperature alone, of the bath, will often cure chronic rheumatism."

Dr Spencer Wells says that "he has treated by it, with great success, cases of gout, rheumatism, neuralgia, and skin diseases, affections of the kidney, dropsy, paralysis of the lower limbs," &c.

Dr Haughton says, "The medical practitioner will find it an invaluable addition to other treatment in all cases of blood poisoning—whether by uric acid, as in gout and gravel; by the same or by lactic acid, as in rheumatism; by the hydrocarbons, as in bilious diarrhœa; as also in those diseases which arise from a deficiency of nature's purifier—oxygen gas."

Sir Benjamin Brodie says, "The hot-air baths are of great use in dyspeptic and gouty habits, and in those who lead inactive lives." And Baron Alderson says, "I had Sir B. Brodie to see me for sciatica; by his order I have been stewed alive in a bath of 140 degrees—not unpleasant after all." And in an article in the "Irish Quarterly Review," 1858, it is said. "We ourselves have seen obstinate cases of sciatica, which for several years had baffled all the remedies of the most eminent allopathic physicians, yield completely to the benign influence of the Turkish Bath in the course of six weeks. We have witnessed similar effects produced in cases of rheumatism, and contracted joints arising from rheumatic gout; whilst in cases of skin disease it is a sovereign remedy, unrivalled by any other mode of treatment, not excepting the Harrowgate waters. And it should be remembered, that all the beneficial effects here mentioned are experienced, not at the cost of a weakened and debilitated constitution, too often the result of allopathic treatment, but in conjunction with an improved state of health and body, the whole system being strengthened and invigorated, whilst the special disease is driven out." And Dr Rayner says, " In the various forms of neuralgia, and especially in that distressing variety of it called tic douloureux, I have observed the most gratifying results from the thermae."

As to the manner in which the bath effects a cure in this class of diseases, it may be proper to remind the reader of what has been previously stated of the eliminating functions and vicarious action of the skin, and of the vast numbers of pores opening upon it, namely, between seven and eight millions. These innumerable little ducts, if placed in one straight line, would extend over 28 miles. Now, throughout this extensive system of drainage, there should distil every day on an average more than 21 ounces of fluid, holding in solution a considerable proportion of solid matters in a state of incipient decomposition, and also, under the influence of a high temperature, urea and uric acid. In reference to this, Dr Carpenter says, in his Manual of Physiology, that " certain forms of rheumatism are characterised by copious acid perspirations, and instead of endeavouring to check these we should rather encourage them, as the best means of freeing the blood from its undue accumulation of lactic acid ; and it is recorded that in the sweating sickness which spread throughout Europe in the sixteenth century, no remedies seemed of any avail but diaphoretics, which, aiding the powers of nature, concurred with them to free the blood of its morbific matter."

Now the hot-air bath is the most powerful of all diaphoretics. It excites the action of the skin, and thus relieves the liver and kidneys, and other internal organs. It enables the skin to take on their function of eliminating and discharging the accumulated poisonous matter in the blood and tissues. Now gout, rheumatism, sciatica, and other kindred diseases, we know to depend upon an excess of one or other of these acids in the blood, uric acid, or lithic acid, and lactic acid. But the hot-air bath comes in here to our aid as a powerful eliminator of these noxious matters. Funke, under a sweating process, collected in one hour seven ounces of perspiration, which contained $7\frac{1}{2}$ grains of urea—at the rate of 180 grains a day. In another experiment he obtained $9\frac{1}{2}$ grains of urea in an hour of perspiration, being at the rate of nearly half an ounce of urea per day. Uric acid was also obtained, but in smaller quantity. I am indebted to the kindness of Professor Henderson of this University for the preceding statement of Funke's remarkable observations; and after furnishing it, the Professor adds, "You might introduce the foregoing particulars in descanting on the action of the skin in purifying the blood. You might also observe that doubtless many other effete and noxious matters are discharged by perspiration, which the chemist has not yet turned his attention to."

"Surely then," says Dr Cummins, "it is only for us

to give that impulse to the skin excretion which the Turkish Bath is capable of, to make it a powerful means of obviating the dangers of these diseases. What is the chief danger in the various forms of Bright's disease of the kidneys? Is it not non-elimination of urea—that condition of blood called uræmia? But will not the skin, under the influence of heat, eliminate urea, and thus, in preventing further toxæmia, prevent also the apoplexy which depends upon it?" And to the same effect Dr Rayner says, "Congestions of internal organs, especially of the liver and kidneys, are generally soon relieved by the bath; and there is no doubt that a considerable amount of the benefit derived is due to a portion of the blood being drawn from the interior of the body and fixed in the skin, the capillaries of which are particularly large and abundant. The suppression of the excretory functions of the skin is the most frequent cause of structural disease of the kidneys; and Dr Osborn, who has paid considerable attention to the subject, states that out of 36 cases of Bright's disease, 32 arose from suppressed perspiration; and Dr Christison, in his treatise on that disease, says that most of his patients attributed their malady to the above cause. In the paper previously noticed, Dr Goulden mentions that some cases of Bright's disease which he saw, when investigating the merits of the Turkish Bath, were more relieved by the perspiration than by anything he had ever seen."

The same author says,—" In gout and rheumatism, the bath is the best remedy with which I am acquainted, as might be expected, when we reflect that the skin is the organ which is destined by nature to eliminate the poisons upon which these diseases depend. I have a patient, a gentleman upwards of seventy years of age, now under my care, in whom the Turkish Bath cured a most savage fit of gout in the short space of three or four days, although previous attacks (he had been a martyr to the disease more than forty years) had almost invariably continued a month or six weeks, and had always terminated in cold, acrid perspirations, occurring for eight or ten nights in succession; but in the last attack the acrid matter was much more speedily eliminated, and the cold clammy perspiration was exchanged for a warm agreeable one. This gentleman's case was also remarkable, from the fact that he had been suffering from disagreeable head symptoms, as headache, giddiness, singing in the ears, dimness and unsteadiness of sight, which also diminished daily, and finally disappeared."

Thus we see that the whole of this class of diseases are not only preventible but curable by the bath,—by draining off the soluble refuse matter and effete accumulations with which the nervous system has been oppressed, and by rendering the insoluble salts soluble, through supplying them with oxygen. "The use of the bath," says Dr Cummins, "during convalescence from acute rheumatism will generally have the effect of preventing the disease from degenerating into that chronic inveterate form which has always been the opprobrium of medicine; while, for the treatment of the chronic stage itself, there is, perhaps, no other single remedy at all equal to it."

In proceeding next to show the beneficial effects of

the bath in diseases of the lungs and respiratory organs, I cannot do better than introduce the subject with the following letter addressed to myself:—" *March* 10, 1863.—DEAR SIR,—I have much pleasure in bearing my testimony to the efficiency of the Turkish Bath as a cure, in cases such as mine, which I shall briefly describe. You are aware my business is of a sedentary kind, and that I am a good deal confined within doors, but towards the end of 1862 I was more than usually confined; for six weeks I was not out of doors at all; hence my appetite began to fail. However, about the beginning of the year, I was obliged on one occasion to go out on business, and was overtaken on my way home by rather a heavy shower of rain, and caught cold. Next day I felt rather unwell, but nothing to speak of; towards night, however, I began to cough a good deal; throat got sore; I got hoarse; and the strain of coughing brought on a pain in the right side of my chest, which got worse and worse as the night advanced. By the application of mustard (just one poultice after another), early in the morning the acute pain gave way, and I got a little rest. Next day I felt rather shaky— the cough no better; still I thought by taking great care and keeping myself well wrapt up, keeping within doors, &c., I would be well again in a few days. How greatly I was mistaken! Instead of that I became weaker every day, my appetite was quite gone, I could not take for breakfast a quarter of a common halfpenny roll—indeed I could eat nothing. Then came cold sweats during the night: I grew dull and desponding; weakness increased until I really felt it fatiguing to go from one room to another; in short, I have no hesitation in saying, from the state of my entire system, that some deadly disease was being rapidly developed, and that in a week or two more I should have been beyond the power of all human aid. In that state I visited your establishment, and by your advice took a Turkish Bath that same evening. The effect was to me most wonderful. From that moment I have had no cold sweats; the very first Turkish Bath entirely put a stop to them, and by the time I had taken two or three more, my appetite began to improve; that feeling of despondency had gone; I no longer looked only at the blackest side of every thing, but felt comfortable and cheerful—so much steadier both in hand and head that I could go through nearly double the work that I had done for a long time previously. I have now had a course of ten baths, and find my appetite so much improved, that I take for breakfast, not a quarter of a roll, which I could not manage before, but a whole one, and an egg, and two large cups of coffee; beef-tea and bread in the forenoon, and for dinner broth and beef, or roast meat or fish, or indeed anything, although not much at a time. Now, before I took the baths, the smell of either of those dishes sickened me, and to attempt to chew a bit of beef caused me to vomit at once. Such has been my experience of the Turkish Baths; and although my cough is not quite away, my general health is so very much improved, that I have at least strength to cough; and I have no doubt that by their continuance now and again, I shall get rid of that too." Since the above was written, this gentleman has continued his attendance at the baths with such success, that his cough has entirely disappeared. So greatly is he satisfied of the beneficial effects of the Turkish Bath, that he now regularly sends his children to it every week, with marked improvement to their general health; and he states, that whereas formerly they invariably began to pine and fall away as the summer months set in, and cried out to be sent into the country, this year they are quite contented to remain at home and come to the baths.

That consumption is curable, or ought to be curable, has long been the prevailing impression of many of the most eminent of our medical authorities ; and that diseases of the chest generally may be successfully assailed by other means than those which have been hitherto tried and failed, is now becoming a common opinion. The former practice of sending consumptive patients to hot climates is now beginning to be exploded and abandoned. In cases where this disease has been relieved by residence abroad, the benefit derived must be attributed to the action on the skin produced by the heat of the foreign climate. But there is no necessity for ordering patients abroad, when they can have higher temperatures at home, and when similar beneficial results can be obtained much more easily and effectually, and at far less expense to their physical strength and pecuniary circumstances ; the use of the Turkish Bath conferring all the benefits of increased temperature, followed by the tonic effects of cool air, by which the debilitating results of continued residence in a warm climate are obviated.

The main cause of consumption is to be found in an insufficient supply of oxygen to the system. "Close bed-rooms," says Dr Hall, "make the graves of multitudes." Impure air causes impure blood, and impure blood is the origin of consumption. Now, as regards the formerly fashionable remedy of going abroad, are we likely to get more oxygen supplied to us abroad than at home? And when we consider that in the hot-air bath we can get a temperature to suit each particular case, and thus create temporarily the climate we want, and, at the same time, remove in an hour or two more impurities from the system than could be removed in a residence of as many months in a warm climate, and, through the same medium—the opened pores of the skin—introduce plentiful supplies of health-giving oxygen,—when all this can be done at home, what is the use of sending patients to Nice, where, as Dr Barter says, " more native persons die of consumption than in any English town of equal population,"—to Madeira, where no local disease is more prevalent than consumption,—to Malta, where one-third of the deaths amongst our troops are caused by consumption,—to Naples, whose hospitals record a mortality from consumption of one in two and one-third of the patients,—or, finally, to Florence, where pneumonia is said to be marked by a suffocating character, and a rapid progress towards its final stage? Sir James Clarke has assailed with much force the doctrine, that change of climate is beneficial in cases of consumption. M. Carriere, a French physician, has written strongly against it. Dr Burgess, an eminent Scotch physician, also contends that climate has little or nothing to do with the cure of consumption, and that if it had, the curative effects would be produced through the skin and not the lungs, by opening the pores, and promoting a better aëration of the blood."

To all this let me add the powerful and convincing reasoning of Dr Toulmin, in support of favouring the functions of the skin by the hot-air bath, in cases of pulmonary consumption and other complaints :—" It is remarkable how Nature appears to cry out for the use of the hot-air bath : witness the rigors, the night perspirations, which so often succeed to a cold, or, in other words, to obstructed perspiration, and which is evidently the exit which Nature has provided for the escape of morbid matter. But perhaps it is the acid perspiration, so well known from its peculiar and offensive odour, and not confined to phthisical patients (although generally indicating a consumptive tendency), that speaks louder than any other symptom as to Nature's requirements ; and there is none other that exemplifies more clearly the invaluable operation of the hot-air bath in withdrawing from the system the morbid matter (lactic acid) that occasions it, and thus enabling the constitution to re-establish the assimilative functions, which are always more or less deranged during its existence." Dr Erasmus Wilson also says, " If consumption is to be cured, the thermæ is the remedy from which I should anticipate the best chance of success." And Dr Spencer Wells says, " In cases where the lungs or the bronchial mucous membrane are diseased, and the heat and nutrition of the body are suffering from defective arterialisation of the blood, the skin may thus become a compensating organ for the faulty lungs. I was lately informed by a gentleman recently returned from Australia, that a relative of his, undoubtedly in an advanced stage of consumption, had recovered to a most extraordinary degree under the use of the bath and a life in the open air."

I have in my eye at this moment a clergyman who was troubled with weakness of the chest, frequent colds, and constant sore throat, difficulty of breathing, and panting, especially after mounting stairs, together with occasional expectoration of blood and purulent matter from the chest. After two months' constant attendance weekly at the baths, all these symptoms disappeared. The sputa presented no longer a pus-like appearance, but assumed a snowy and crystalline purity of aspect, showing the entirely healthy nature of the bronchial discharge. His chest became stronger, his voice firmer, and his general health vastly improved; while the tendency to colds diminished and disappeared.

On this point, Dr Goulden says of the bath, " Common catarrh is cured at once, and the weekly use of the bath diminishes, or altogether suspends, liability to the attacks. This includes what are called colds in the head, quinsy, sore throats before suppuration, the common winter coughs, and some forms of diarrhœa. Pains in the muscles, after excessive and unusual fatigue, are removed; and those who suffer pains in the seats of old injuries, upon slight changes of the weather, lose these pains entirely." Dr Culverwell says, " Where the Turkish Bath is regularly indulged in, the skin always maintains sufficient vitality of its own to resist all changes of temperature, whereby the bath becomes a certain preservative from all colds, coughs, and their distressing consequences—bronchitis, consumption, &c." And Archdeacon Goold, in his narrative of a visit to the bath at Cairo, says, " I dressed myself and walked home, leaving my enemy, the ague, hovering, I presume, within the precincts of the bath."

I shall conclude this part of the subject with the following letter addressed to myself:—" Sir—As I am about to leave town, and not likely to return for some time, I have felt that I owe you some thanks for the very kind and courteous manner in which I was treated by you at your establishment. Feeling also a great renovation of my health and strength from the time I commenced making use of the Turkish Bath, I feel convinced that it is to that agency that I must attribute my recovery. You are aware that at the time I consulted you, my case presented a rather peculiar aspect, and I would strongly recommend any one labouring under a similar affection of the chest to try the medium of the hot-air bath. I was at first inclined to mistrust its efficacy, but by patiently continuing I have felt a gradual improvement—so much so, that I trust my complaint is almost eradicated. Should you ever require a testimonial intended for publicity, I will feel much pleasure in explaining the benefit I have received under your treatment." On the above case I need only remark, that there was considerable dulness, on percussion of the chest, over the upper part of the right lung, which had existed for some time, so much so that I apprehended incipient condensation of the lung.

The effect of the Turkish Bath was entirely to remove that dulness, and also other complaints attendant upon it.

Again, in febrile diseases, Dr Cummins considers that the eliminating influence of the Turkish Bath may be equal to expelling the poison from the blood, especially in the first stages, and so of cutting short the attack. In the "Cork Examiner," it was stated lately that some medical men in Cork require the midwives to take the hot-air bath to destroy puerperal poisons, and that many a mother might be spared to her family by this precaution. Dr Cummins also considers the hot-air bath the most powerful means at our disposal in the treatment of dropsy. In all skin diseases and cutaneous eruptions, it is the only remedy at all adequate to the complaint. These are only efforts of nature to expel a constitutional taint. Dr Tucker thinks, therefore, that people would not be disfigured with blotched face, or any form of eruption of the skin, if they would only take the hot-air Roman Bath. All scrofulous habits of body, tending to, and often ending in, pulmonary consumption, will, I have no hesitation in saying, have their abnormal diathesis corrected, and their hereditary predisposition to phthisis and other diseases checked and removed by the persistent use of the Turkish Bath. "In blood diseases," says Dr Brereton, "I see no necessity to put any limit to its curative powers. Diseases of the skin are, for the most part, mere efforts, more or less ineffectual, to throw off impurities, and are, of course, superseded by the bath. This we know, that scrofula and gout are almost unknown in countries where the bath is in vogue. With all these facts, is it not strange that we should so long have neglected it?" Seeing, then, that it is chiefly the presence of this concentrated essence of all poison, the scrofulous taint, that gives fatal malignity to scarlet fever, measles, hooping-cough, and a variety of other diseases, for all of which the hot-air bath is recommended as a powerful and specific remedy, it is matter of regret that so much ignorance and apathy should prevail respecting one of the most potent remedial agents with which we are acquainted.

The following case of diseased liver, with severe ulceration in the throat, may be perused with interest. Captain ———, aged forty, of a very dark, sallow complexion, and with great anxiety depicted on his countenance, states that he has been for several years in India and China; suffered much from liver complaint, and had taken large quantities of mercury, and been much worse ever since his return to this country; complains of severe pain on the right side, over the region of the liver, extending through the chest to the left shoulder; is also troubled with a large ulcer of a mercurial kind in the throat, immediately behind the uvula. His medical attendant was under the impression that some portion of bone was involved in the disease, and that the sore would not heal until the separation of the diseased bone. The appetite was bad, the digestion wretched; sleep almost entirely gone; great debility, depression of spirits, and despair of getting better; had tried various kinds of baths in London without benefit. The treatment commenced with the chemical, galvanic, detergent, and occasional Turkish baths, and was steadily persevered with for six weeks. During the course of treatment, at the first all the symptoms increased in intensity, but afterwards began rapidly to subside. The result was most satisfactory. The health and vigour of the body were wonderfully restored. The dark, sallow, and dismal appearance of the countenance vanished; the appetite became vigorous, the sleep perfectly natural, and the glow of cheerfulness and delight beamed in the countenance. The ulcer was in a short time entirely healed, and the gentleman left in the most excellent health—a blessing to which, for some years previous, he had been a stranger. This case is a very striking illustration of the very great efficacy of the various Medicated and Turkish Baths in the rapid cure of many formidable diseases which had for years resisted every other mode of medical treatment. It was gratifying to observe the rapid progress in health, in proportion as that pernicious drug, mercury, of which he had taken a large quantity, was being deterged from his system through the pores of the skin. The cure in this case has been permanent and effectual, no relapse having occurred up to this time, although it is twelve months since the treatment was completed.

I do not mean to assert that the hot-air bath, or any other bath, is a panacea for all the ills that flesh is heir to; but I will assert, and I do it advisedly, both from my own observation and experience and from the testimony of many eminent authorities in the medical profession, that it is better fitted than any other means to be a remedy in most cases of curable disease, and that it is calculated to be of great benefit in many which have been considered as hopelessly incurable. Dr Brereton says, "In short, it has become a question with me, not what the bath will cure, but what it will not cure." "If you want to live long and healthy, I again say, take the bath: if you want to save doctors' bills, take the bath. I do not mean to say that it altogether supersedes medicine, but I do say that in a vast majority of, if not in all, blood diseases, the bath is more speedy, more certain, and far more agreeable, than any other treatment of disease." Dr Cummins says, "The bath is one of those agents which are found by experience to have a wide range of application in the treatment of disease; and there are several such, and no physician thinks of condemning them because they are such." "The Turkish Bath is powerful in disease, and for that very reason the Turkish Bath should be regularly prescribed, and its effects watched by the physician." And Dr Madden says, "It always appeared to me, as it does now, that the introduction and use of the Turkish Bath into these countries, as it really existed in the East, were it under judicious, legitimate medical control and direction, would be a very great blessing to humanity, and a most important adjunct to medical skill and science. There are very few diseases, in my opinion, that might not be benefited by the use of the Hamâm." And, finally, Dr Haughton says, "I am persuaded that it has a wide range of application, and that there is no class of society which would not be benefited by its introduction. I do not advocate a panacea, but I recommend an institution which will prevent as well as cure disease; which comes down to us from the most remote ages, and is now used by a large proportion of the human race; which is venerable from its antiquity, founded upon science, supported by authority, and confirmed by experience."

I might, were it at all necessary, quote other authorities equally favourable, and all bearing strong testimony to the efficacy of this bath as a most potent agent, combined with hydropathy, in the cure of many diseases not easily cured by other means. And I have no hesitation in saying that the resuscitated Oriental bath is the very thing we want, and is better fitted than any other means to arrest the progress of many of those maladies which we all deplore. This desirable result is obtained, as we have seen, by acting on the skin, eliminating from its millions of pores and multitudinous cells effete and poisonous matters which are generated in the system, and which, if proper attention is paid to the functions of this very important organ, are carried off in the form of insensible perspiration, but which, if obstructed in their course, or re-absorbed into the blood as irritating poisons, become the fruitful source from whence proceed malignant and fatal maladies.

Before concluding this chapter, I shall refer to a remarkable application of the principle of the hot-air bath to the cure of insanity, in the District Lunatic Asylum, Cork. At a meeting of noblemen and gentlemen, on the occasion of presenting an address and testimonial to Dr Barter, for having introduced the curative system of the hot-air bath into that country—the Lord Lieutenant of the county in the chair—a statement was made by Dr Power, the resident physician of the Lunatic

Asylum there, of the extraordinary results that had attended the introduction of a Turkish Bath into that institution. He said—"The first persons submitted to its influence were much pleased with it, and were anxious to go again. After about four months' use of it, I found that 17 persons had been perfectly cured by it, and sent home to their friends. The cases to which I now allude were a long time in the house, and classified with the incurables. From 50 to 80 patients are daily submitted to its influence; many for its remedial action, but the greater number for motives of cleanliness; even these latter are wonderfully improved in appearance by its use, and have acquired the ruddy glow of health, instead of the pale and sickly look of invalids. Of course, out of more than 500 patients in the institution, all were not expected to recover, nor were they all under treatment for the purpose; but the best way of showing the effects of the bath would be by statistics. It was only fair to conclude, that if the proportion of cures had been greater since the introduction of the Turkish Bath than before it, this bath must have had some influence in producing that desirable result. I see by my notes, that for the year ending March 1861, the cures were 59 per cent.; but for the nine months ending 31st December last, during which period the bath had been in use, the percentage of cures was 76—that is, 74 had been cured out of 96 entered. That was more than double the number of cures produced in any asylum in England.*

* Dr Power adds, in a footnote in the printed document, and in his own handwriting, the following observation:—" When the bath

The patients, after the first few baths, all seemed to be much pleased with it, and were always longing for the time when it was to be administered. Those who had suffered a relapse after having been sent out cured, showed no unwillingness to return to the asylum; and even asked to be taken there at once, in order that they might get the bath, as they considered that nothing else would cure them. I have never seen any ill effects from the bath, except a little nausea and a slight fainting in a few instances, but after a bath or two those effects disappeared. Up to that time I have used it in more than 900 cases; and since March 1861, 30 idiotic patients have been removed to a higher class, and rendered capable of enjoyment, and of doing work about the establishment. I would recommend the introduction of the Turkish Bath into all public institutions, and I am firmly convinced that it has as beneficial an influence on the system as air and exercise."

I have before me an extract from a report of a lecture delivered in Mary's Chapel, by Dr Yellowlees of the Royal Edinburgh Asylum, Morningside, in which the lecturer speaks of brain diseases, and the various causes from which they proceed—namely, extreme emotion or agitation, over-working or over-straining the brain, intemperance, vice and immorality, and hereditary transmission. The doctor adverts also to the chief mental medicines relied on for its successful treatment—namely,

was not in being in 1860, the deaths were 48. In the nine months ending December 1861 they were 18, or at the rate of 24 in the year—i.e., exactly one-half the number."

something to do, something to enjoy, and somebody to trust; among which are included games of all sorts, music, literature, regular balls and concerts, and a literary journal conducted by the patients themselves. I have no doubt that the treatment is based upon sound, and humane, and scientific principles. Still there is something wanting to make it complete—and that something the doctor does not even name among the list of remedies which he enumerates. Either he is ignorant of its virtues, or, if cognisant of the facts related above, he ignores it as unworthy of notice in his lecture; while others, equally if not better qualified to judge, have proved that the hot-air bath is most powerful as a curative agent in many cases of insanity, and most exactly fulfils all the conditions required,—something to do, something to enjoy, and something to trust. Is it not surprising that, after the efficacy of this most precious remedial agent has been again and again demonstrated by many astounding facts, there should be so much apathy, and such a want of public spirit on the part of the directors and managers of our own Royal Infirmary, Lunatic Asylums, and similar institutions? In the bath erected by Sir John Fife, in connection with the Newcastle Infirmary, more than 1200 patients have been treated, with truly wonderful and most beneficial results in the great majority of cases. Doubtless, if there were a real appreciation of the intrinsic merits of this valuable boon and powerful curative agent, especially by the medical officers of such noble institutions, the thing desiderated would soon be an accomplished fact. May the time soon come when that which has already proved itself so great a blessing to many, will be recognised in all our public institutions,—be placed within the reach of the community at large, and be prized by all classes of the public, as at once a luxurious indulgence, a purifier of the blood, a preservative of health, and a remedy against disease.

CHAPTER VIII.

OBJECTIONS ANSWERED.

NOTWITHSTANDING the utility and advantage of the Turkish Bath, as a most powerful curative agent, and an invaluable conservator of that precious boon HEALTH, there are many popular prejudices entertained against it. I now proceed, therefore, to answer the objections that are most frequently urged against its adoption and use amongst ourselves.

1. Among the most important and common objections to this bath, is—*The supposed danger of catching cold after it.* This cold-catching idea is almost universal among all classes of the community, at least in this country. But I trust I shall be able satisfactorily to demonstrate the fallacy and utter groundlessness of this absurd notion. It is like the baseless fabric of a vision. It rests upon a foundation of sand, and a mere touch will throw it to the ground. This mistaken supposition has evidently arisen from the fact that many have really caught severe colds, owing to their own imprudence, after taking other baths—common warm or cold baths. And this may arise from either having recourse to them at improper times, or in that condition of the bodily health in which such a remedy is contra-indicated;

and also from the fact that few persons seem to be aware of the usual precautions necessary to prevent this distressing result. Feeling themselves a little out of order, they think that a common warm bath would be of service in quickly restoring them to their ordinary health, but they frequently lose the benefit, in consequence of not having a cold shower or pale douche poured over the body after it, and also smart friction while drying the surface of the body; or it may arise from undue exposure to cold, thereby chilling the surface, and causing the blood to be driven in upon the internal parts, producing inflammations, colds, and other severe affections.

In order to show the bad effects of injudicious bathing, first, in common warm baths, I will quote the following description from Sir Arthur Clark, M.D., of the striking sensations experienced on entering a warm bath heated above 98 degrees. He says, " the pulse is accelerated in frequency and force, the superficial veins become turgid, the face flushed, the respiration quicker than natural, and sometimes laborious; and, if the immersion be continued beyond a given time, determination of blood to the head is greatly increased, the arteries of the neck and temples throb violently, a sensation of anxiety at the heart comes on, threatening suffocation, the person grows giddy, and feels a fluttering at the heart, and if these warnings are not attended to, the bather soon becomes insensible, and is carried off by apoplexy." We frequently hear of persons dying in a warm bath, doubtless from this very cause, the bath being above 98 degrees, that is, above blood heat. But such things are not likely to occur in baths where there is medical superintendence, and not at all in *Turkish* Baths, or *Hot-Air* Baths, heated to even a far higher temperature. Such are not at all the sensations experienced in the Turkish Bath when properly administered.

Again, bad effects often arise from a common bath, either hot or cold, being taken too soon after a meal. It is well always to remember that an interval of two hours at least should be allowed to pass before taking any bath after taking food. And I may even observe that, after all, the advantages derivable from these common baths as curative agents are very doubtful. They may be very good for promoting cleanliness, by removing perspirable matter from the surface of the skin, but they will never be equal to the Turkish Baths, even in this particular. The common warm bath is imperfect in various respects. In the first place, the bather is put at once into a high temperature, and cannot pass from grade to grade of heat as in the Turkish Bath. Besides, as a learned writer has observed, it is not natural for a man to be immersed in water; it might do well for a fish, but it is not suited to man. The highest temperature of water that can be borne is about 105 degrees, but the bather is obliged to come out of the water in about fifteen or twenty minutes from sheer distress. The pressure of the dense medium about the body prevents perspiration, and no oxygen can enter the system during the period of immersion.

The same remarks are applicable to cold and sea-bathing. I could mention several cases which came under my own observation. I will only notice one. A strong healthy young man, about eighteen years of age, who had just completed his apprenticeship, had invited some of his companions to spend a day with him at his house, which was in immediate proximity to the sea. *After dinner* they proposed to bathe together in the sea, which they did. They all escaped with impunity except this poor young man, who was seized with symptoms of apoplexy while in the water, and was carried home and died that night. Now this melancholy result was caused by imprudently venturing into the sea at an improper time, immediately after dinner, and I have no doubt that if it had been a warm bath instead of a cold one, the result would have been equally disastrous.

I mention these facts to show that much of the prejudice existing against the Turkish Bath has been in reality derived from common baths, and transferred to it from them—and from the injurious effects produced under certain conditions by them. The former, the Turkish Bath, is safe, salutary, and beneficial; the latter, the common baths, are not nearly so much so even under the best conditions, and are, in very many cases, taken in circumstances calculated to produce very pernicious results. The late Professor Alison told me one day, when visiting a patient in consultation, the case being one of inflammation, that they had generally a greater number of such cases in the Royal Infirmary during the sea-bathing season than at any other period of the year. And Dr Clark observes, that the predilection which mothers have in general for cold bathing, in order to brace and strengthen their children, as they conceive. accounts for the frequent cases of hydrocephalus, or water in the brain, which occur in the early periods of life, and also of bilious and liver complaints in later life, which are decidedly the most prevailing disorders in Dublin. These evils, he considers, arise, in a large measure, from the indiscriminate use of the cold bath.

I might go on at greater length, but it is enough for my purpose, in so far clearing my way, that I have demonstrated and established the fact, that the indiscriminate use of common warm and cold bathing is a fruitful source of severe colds and dangerous inflammations. The odium which it is attempted to fasten upon Turkish Baths in the infancy of their introduction amongst ourselves, has arisen entirely from the acknowledged ill effects of injudicious common bathing. Is it not unjust, then, to transfer to the innocent the crimes of the guilty?

To come now to the point: Methinks I hear some one ask—" Do not the same remarks apply to Turkish Baths? Is there not as much danger of catching cold from them as from the other baths of which you have been speaking?" I unhesitatingly answer, *decidedly not*. And I will endeavour to prove that there is no such danger to be apprehended from this bath. In so far as my own experience goes, both as it regards myself and thousands of others who have availed themselves of the Turkish Baths with which I am connected, not one of them, to the best of my knowledge, has ever been so affected. Hundreds have been speedily cured of severe colds and other diseases, from which they had suffered for weeks and months before they came to the bath. But not even in one instance has a cold been superinduced. In fact, those who frequently use the Turkish Bath do not know the way to catch colds, and with ordinary care they could not catch cold even if they were willing. I shall here mention a curious case. One gentleman, who has been a "Companion of the Bath" since the time that the Sciennes Hill establishment was opened, and who had for years suffered much from a deranged condition of the digestive organs, with occasional severe headache, obstinate constipation, and susceptibility to colds, and was frequently confined to bed and obliged to swallow large quantities of blue pill and other nauseous drugs, since he commenced taking the much-abused Turkish Baths has never required even one pill, or a single dose of any kind of medicine, nor has he, up to the present time, been troubled with headaches, and his appetite is now vigorous and first-rate. This gentleman lives a few miles from town on his own estate. He possesses a very choice selection of rare poultry, and some parties, who seemed to fancy they had a better right to them than their owner, thought proper to make so free with them as to abstract them from the premises during the silent watches of the night, forgetting that although the eye of the owner could not see them, there was another eye gazing on them to which the darkness is as the day,—the eye of Him who will bring every secret action into judgment. The gentleman was very much annoyed at losing his favourite chickens, and resolved to watch all night along with a police-officer; and he actually lay upon the damp grass, and in the midst of a heavy rain, and was not a bit the worse of it. And he says himself that if he had done such a thing previous to his commencing the Turkish Bath, the consequences would have been most serious. But, after all, this is no great thing; it is only the natural result of the bath, the tendency of which is to fortify the system against the pernicious influences of atmospherical vicissitudes.

As an illustration of this truth, that the Turkish Bath has the effect of hardening and fortifying the skin, so as to render it almost insusceptible to the influence of cold, a remarkable and amusing fact is related by Dr Erasmus Wilson as having occurred in the experience of Mr Urquhart. "A fine athletic child," he says, " of five years old has been brought up in the bath, and has never worn other clothes than a loose linen garment. He is a sturdy little fellow, with the independence of deportment of an Indian, and the symmetry of an Apollo. He was met one wintry day, when the snow was on the ground, walking in the garden, perfectly naked. ' Do you feel cold?' inquired his interlocutor. 'Cold!' said the boy, touching his skin, doubtfully, with his finger, ' yes, I think I do feel cold.' That is, he felt cold to his outward touch, but not to his inward sensations, and it required that he should pass his finger over the surface of his body as he would have done over a marble statue, to be sure, not that he was cold, for that he was not, but to be convinced that his surface felt cold." He also relates another case, proving that the bath gives endurance, and that that endurance fortifies a person against a very prevalent cause of discase in this climate, namely, colds and affections of the chest. It is the case of a clergyman who was scarcely ever free from colds during the winter-time, and these colds were often so severe as to lay him up for several weeks. He also suffered from attacks of neuralgia. But after adopting the use of the bath twice a week, all disposition to colds and neuralgia ceased, and for the first time in sixteen years he passed the winter without a cold.

As further disproving this unfounded notion, be it only remembered that the Russian rolls in the snow after the bath, not only with impunity but with delight. To disabuse the minds of the inhabitants of this country of the idea that the Turkish Bath exposes to cold, or is dangerous in cold or winter weather, let it only be remembered that the Turk himself is obliged by his religion to go to the bath regularly both summer and winter, without respect to weather or season. This bath, in fact, actually renders cold weather innocuous. Count Rumford made the experiment, by exposing himself to cold after bathing, for the purpose of ascertaining this, and found the effect quite the contrary. He could endure every kind of inclemency, which before would have laid him up. And Dr Haughton of Dublin says that he once involuntarily experienced the same thing when in Constantinople, where sudden changes of weather and temperature are exceedingly common. "Having gone up the Bosphorus a long way, to a bath unfrequented by Europeans, I found on going out that a sultry morning had been succeeded by a wintry afternoon, and I had no way of returning but by an open boat or caïque, in which I got thoroughly chilled, but did not take cold."*

There is much popular error on this subject of taking cold, notwithstanding all that has been written on it. It is supposed that when the body is warm, it can worst support a rapid abstraction of caloric. But, however true this may be when the heat has been obtained by bodily labour, severe physical exertion, and at the expense of nervous energy, it is far from being true of that heat which is communicated to us from without, and which is, in truth, an available surplus, which we can afford to lose. Besides this, after a bath the capillary circulation is more vigorous, and we are therefore in a better condition for resisting cold than before. Thus, Dr John Le Gay Brereton, in a lecture at Bradford, on the Action and Uses of the Turkish Bath, says, "After leaving the hot room in our Bradford Bath, bathers were in the habit last winter of jumping into a bed of snow, which had been collected for the purpose. I have myself spent the whole night in the woods at Blarney, without any clothing save the bath-sheet, after coming out of Dr Barter's bath at that place. This was after a ball, when, with several other gentlemen, we had retreated to the bath for the sake of refreshment from fatigue. So delightful was the cool air, that, when far away from any dwelling, we threw aside even our sheets to enjoy the morning breeze at daybreak. You need not then fear exposure to the air after the bath. It is not so much for the sake of *cooling* that this process is necessary, as

---

* The Facts and Fallacies of the Turkish Bath Question. By Edward Haughton, M.D., Dublin.

to keep up the action of the bath by exposing the skin to the air. It is to compel the skin to *breathe*, after having put it in a state of ability to do so by cleansing the pores. Now, having taken the bath, you will continue warm, without clothing if you like, until the skin becomes again blocked up."

In far colder countries than ours, in Russia, Lapland, Sweden, Norway, and Denmark, there is no cottage so poor, no hut so destitute, but it possesses its vapour bath, whither all the family resort every Saturday at least, and every day in case of sickness, and in which the inhabitants experience both comfort and salubrity. Even there it makes so necessary a part of the system of living, that it is used by people of every age, in all circumstances, in almost all sicknesses, before and after a journey, after hard work or excessive exercise, to obviate the effects of fatigue; and yet the people of these countries never even dream of catching cold. Nor does the inhabitant of any other country where the bath is a national institution. It is only in this country, where the people readily admit that cleanliness is next to godliness, and yet in practice ignore the very truth which they profess to believe, that this unfounded and erroneous idea is entertained to the prejudice of the hot-air bath.

I trust I have said enough to do away with this misapprehension. The bath produces no such chill, and occasions no such shock to the nervous system, as is produced in the reaction witnessed every day among us, and commonly termed "a cold." Cold arises from checked perspiration, but it is impossible that the perspiration can be checked when the pores have all been freely thrown open; and, as this is what the Turkish Bath does, removing everything that impedes and checks a free perspiration, therefore there is no fear of the bath giving cold. In the bath we perspire freely, and the internal organs are purified. We rinse off the perspiration with warm water, and then gradually shut up the pores with tepid, and then with cold water. The body is thus regenerated and invigorated. We then rest tranquilly until every particle of moisture is removed from the skin. Internal heat is all the while being now more rapidly generated, through the greater absorption of oxygen by the purified pores of the skin. And when we are thoroughly dried we dress and leave the bath—serene, composed, and happy. Where is the danger of cold in all this? It is an entire chimera. There is not a chink now by which cold can enter. With these conditions properly observed, we boldly and triumphantly assert that the bath cannot give cold.

2. The second popular objection which I now proceed to notice is—*That the Turkish Bath is weakening.* I unhesitatingly affirm that it is not. It is the very reverse. The effect of the bath is powerfully tonic and invigorating. This can be proved to a demonstration. Look at the bath attendants, for example, who are from four to five hours daily in the bath at one time, and are not debilitated in the least, but rather invigorated in every way, and absolutely increase in weight. Two of my bath-men gained eleven pounds in four or five months. The only feeling they have when coming out of the bath is that of hunger. This is easily accounted

for, from the fact that the process of digestion is rendered more active, and nourishing material is more quickly supplied from the digestive system to the various tissues of the animal economy, and thus the entire frame is strengthened and invigorated. There is a gentleman attending the baths at present, afflicted with asthma and chronic bronchitis. Indeed, the lungs themselves are affected. He has been suffering from the complaint for two years. The body is very much attenuated, accompanied with great prostration of strength. Up to the present time he has only had eight baths, and is so much improved that he has actually gained one pound in weight, and says himself that he has derived more real benefit from these few baths than from all his previous medical treatment and changes of climate for the last two years.

This idea, that the bath is weakening, has obviously been suggested by the loss of fluids during the process of free perspiration. But never was a greater mistake made. For, first, the abstraction of fluids while perspiring is neither caused nor attended by any violent physical exertion, and therefore cannot be weakening. Dr Carpenter, in his Prize Essay, observes, that perspiration has no weakening effect in itself, except by the diminution of the water in the blood, which may be resupplied from the stomach; and that this appears from the fact, that if persons exposed to a high heat make no bodily exertion, they experience no loss of vigour, if copiously supplied with cold water; but, on the contrary, such exposure will conduce very much to invigorate the system. Secondly, because the perspiration produced drains away no living tissue, but merely effete and poisonous matters, which were oppressing, and not maintaining life. And, thirdly, in exchange for this dead material lost, the bather gains new and increased supplies of oxygen into his system through every opened pore of his body, and oxygen is the quickener of every function, and the very life of the body. So that if he perspires well, he comes out of the bath actually stronger than when he went in. His appetite also is increased, and what he eats and drinks is sure to be more rapidly assimilated, for a demand has been created at every point by the greater activity of function which has been excited. If he even goes into the bath weary and jaded, he comes out of it, not weakened, but, on the contrary, refreshed and strengthened; for such is the exhilarating effect—an effect which it is impossible to resist—of the increased supplies of oxygen which the body now inhales at every point, and imbibes at every pore.

Further, this mistaken idea of loss is not only no loss of materials necessary and salutary to the economy of the human body, but neither is it a loss which can affect the nervous and muscular powers of the individual otherwise than beneficially. For nutrition depends upon the depositing of new material in place of the old. The function of the skin is emunctory. If active, the removal of the old material will be rapid, and in proportion to the energy and completeness of its removal will be the eagerness of the tissues to take up new material from the blood. But if torpid, from the pores being closed up, then this source of nutrition is impeded and frustrated, and the body becomes either atrophied or diseased. What the bath does then is, first of all, to cleanse the surface of the body free of impurities, to soften the callous scarf-skin so that it may easily be peeled off, to remove all the minute scales and particles of dead skin, and then further to give free egress out of the body to all the effete and worn-out material which poisons it, and generates disease within it. The copious perspiration induced by the bath, then, is a loss so far as this is concerned. But what a blessed loss! A loss only of what is hurtful and pernicious. And when the quickened and increased power of rapid and vigorous assimilation from food and drink after the bath is taken into account, then it may be seen at a glance that the loss is a great gain!

As an illustration of how necessary free perspiration is to the due and healthy working of the animal economy, it may be mentioned that in India and China the first question put to a new-comer from this country is, "Do you perspire freely? Because, if so, you will do for this climate, and have good health." There the daily waste is as quickly repaired as it is lost, and those whose skin is in good working order endure the climate for many years, and many actually get fat.

Look at the large smelting-houses in England and Wales, where the stokers are kept for many hours in a deluge of perspiration. The drain is very active, and it is consequently necessary to supply the loss with copious draughts of water. With this object, says Dr E. Wilson, "Each man is allowed a certain quantity of oatmeal daily. The oatmeal is served out by the foreman, and is scalded with hot water and made into thin gruel.

This is the drink with which the men supply the loss of the perspired fluid. They give forth, in the shape of perspiration, water holding in solution the used and useless materials of the frame, and they receive in return a wholesome nutritive material. Can we wonder that these men are perfect athletæ in form, that they are in the finest possible condition for labour ; and that, although working in buildings open to every draught of cold wind, and bathed in perspiration, they never, indeed they cannot, take cold." In the copper-smelting works at Swansea, the heat between the furnaces at which the men work is 200 degrees of Fahrenheit; they drink a gallon of thin gruel every hour, working four hours at a stretch, and the ground on which they stand is a pool of perspiration. And yet, with rivers of perspiration streaming down their athletic frames, such men are healthy, long-lived, and happy. But there is nothing nearly so violent as this in the Turkish Bath.

I might adduce other proofs to do away with the false impression that the bath is weakening. "We can test the point," says Mr Urquhart, "in three ways,— its effects on those debilitated by disease, on those exhausted by fatigue, and on those who are long exposed to it. First, in affections of the lungs and intermittent fever, the bath is invariably had recourse to against the debilitating nightly perspirations. The temperature is kept low—not to increase the action of the heart or the secretions; this danger avoided, its effect is to subdue, by a healthy perspiration in a waking state, the unhealthy one in sleep. No one ever heard of any injury from the bath. The moment a person is ailing he is hurried off to it. Second, after long and severe fatigue—that fatigue such as we never know—successive days and nights on horseback. Having performed long journeys on horseback, even to the extent of ninety-four hours, without taking rest, I know by experience its effects in the extremest cases. A Tartar, having an hour to rest, prefers a bath to sleep. He enters as if drugged with opium, and leaves it, his senses cleared, and his strength restored as much as if he had slept for several hours. This is not to be attributed to the heat or moisture alone, but to the shampooing, which in such cases is of an extraordinary nature. . . . . . Well can I recall the *hamâm* doors which I have entered scarcely able to drag one limb after the other, and from which I have sprung into my saddle again *elastic as a sinew and light as a feather*. . . . . . You will see a *hammal* (porter), a man living only on rice, go out of one of those baths where he has been pouring with that perspiration which we think must prostrate and weaken, and take up his load of five hundredweight, placing it unaided on his back. Third, the shampooers spend eight hours daily in the steam; they undergo great labour there, shampooing perhaps a dozen persons, and are remarkably healthy. They enter the bath at eight years of age; the duties of the younger portion are light, and chiefly outside in the hall to which the bathers retire after the bath: still there they are, from that tender age, exposed to the steam and heat, so as to have their strength broken, if the bath were debilitating. The best shampooer under whose hands I have ever been, was a man whose age was given me as ninety, and who, from eight years of age, had been daily eight hours in the bath. This was at the natural baths of Sophia. I might adduce, in like manner, the sugar-bakers of London, who, in a temperature not less than that of the bath, undergo great fatigue, and are also remarkably healthy." I could mention a number of other facts and cases equally favourable within my own experience, but after the preceding array of evidence this is altogether unnecessary.

3. The next objection to which I shall advert is—*The danger of causing a determination of blood to the head.* This idea is started in the " Glasgow Medical Journal" for July 1862. The writer animadverts in strong terms on the introduction of this most valuable bath into this country, as if it were a novelty in, and unsuited to, this climate, whereby he only exposes his own stupidity and ignorance, as will be afterwards clearly demonstrated. He next asserts that the bath is injurious in its physiological action on the brain and nervous system—which only serves to show his own crass dulness of apprehension, in never having clearly discerned that its true and primary action is on the skin, and then, secondarily, on the other organic functions of the body. And not only is its action on the brain and nervous system *less* than its action on any other part of the animal economy, but whatever action there is on these parts is of a soothing and sedative nature, unless indeed the stomach be overloaded at the time, and then it only shows, what we all already know, that the stomach, and not the bath, acts on the brain.

But he goes on to say, that in his personal experience

it gives rise to " acute frontal headache, with throbbing of the temples, and distinct tendency of blood to the head." I quote the writer's own words, to show that I desire to do full justice to him. And at first sight the charge seems to be serious. But " he that is first in his own cause seemeth just ; but his neighbour cometh and searcheth him." I hesitate not to declare that the charge is altogether unfounded, and that it does not require much penetration to discover the writer's ignorance and want of candour. In the first place, he does not inform us how often he had subjected himself to the influence of the bath, or whether he had given it a fair trial, or of the condition of his own health and digestive apparatus when he went into the "stifling atmosphere"— as he is pleased to term it, of the bath. Hundreds and thousands of his countrymen have discovered that there is nothing stifling about it. Yea, many individuals afflicted with diseases of the chest, for example, have learned to breathe more freely by means of it.

Further, I am quite at a loss to understand what kind of a bath this unfortunate writer had got into. Probably into one of those numerous places fitted up at small expense in wretched cellars, under ground, where the atmosphere must necessarily be disgustingly stifling. A bath heated and filled with warm damp vapour, instead of pure hot air, is probably the kind of thing he has patronised. What does he think of Lord Carlisle, the present Lord-Lieutenant of Ireland, and the late Archbishop Whately meeting together in a Turkish Bath in Dublin —as I know for a fact they did,—and enjoying it? Could they, with the crystal beads of perspiration, like limpid dewdrops, begemming their foreheads, yea covering their whole bodies, and, like a couple of monks of old " devoutly counting their beads,"—have been such devoted and determined C.B.s (Companions of the Bath) if they had not liked it, and been of an entirely contrary opinion to the Glasgow editor? And when a gentleman who was present with them in the bath was asked how the Lord-Lieutenant liked it, he replied, " Why, he just perspires like any other man."

To give my own experience on this point, first of all, I have been in baths in Glasgow and other places, and also in my own baths innumerable times, and have never felt the atmosphere stifling in the least, nor have I ever experienced either frontal or lateral headache at all. And further, thousands of persons, both ladies and gentlemen, have under my own superintendence passed through the bath, and not one, to the best of my knowledge, has ever exhibited even an approach to the violent and peculiar symptoms which this Glasgow editor complains of. It is true enough that in a few cases, and on taking the first bath, but far less frequently on arriving at the second bath, there is occasionally a slight degree of headache, but nothing like the acute and violent symptoms of which he speaks. And even in such rare cases it generally arises from a deranged condition of the digestive organs, or from the morbid material in the system being poured into the stomach and intestines. And with subsequent proper treatment it quickly passes away. But even admitting the occurrence of uneasy sensations at first, is it not worth while enduring a little for the sake of a great and certain good? And is it not like the impatience of a schoolboy, to decry the whole system because in his apprenticeship he does not reap all its benefits, and experiences at first a little inconvenience?

From my own experience then, as well as from the testimony of many thousands who have entered the baths under my superintendence, I confidently affirm that such symptoms as the Glasgow editor describes cannot possibly occur in a well-regulated bath, and where the bather is not allowed to remain too long in it the first time, or unless owing to the violation of some well-known dietetic or sanitary law. And, indeed, there is, after all, something so very peculiar in the symptoms described, that one is naturally led to the conclusion that the worthy editor's experiment must have been post-prandial. I have invariably found that where such symptoms appeared at all, they were, in almost every instance, the result of indulgence in the luxuries of the table shortly before entering the bath. What other result can be looked for, if persons offer to enter the bath immediately after dinner—which I, however, do not allow them to do—and even, as some of the parties referred to have frankly confessed to me, after taking more than one tumbler of such intoxicating beverages as toddy, or Glasgow punch? Should any Glasgow citizen, or even a Glasgow editor, venture into the Turkish Bath, or even into any kind of bath after that, what else can be expected than headache, and throbbing of the temples? It must be apparent to every one, even to the worthy editor himself, on calm reflection, that in such cases the blame is not attachable to the Turkish Bath but to the parties themselves.

The learned editor is apparently much interested in the safety and stability of insurance companies, for it is in their interest avowedly that he is writing, strange to say, in a medical journal, and not in the interests of truth and science, which would have led him candidly, dispassionately, and without prejudice to consider the whole subject, and weigh well his statements. He wonders, therefore, that Dr Fleming, in his recent work on Life Assurance, should have passed over in silence this great fact, which the learned reviewer has only just discovered, namely, that the Turkish Bath shortens the duration of human life, thereby involving such institutions in a much greater amount of risk than formerly, when there were no Turkish Baths; and in order to prove to a demonstration the truth of his statement, states, that " one of the leading insurance companies had a death among their insured." Very likely, indeed! Who ever supposed that Turkish Baths, even with the help of Insurance Companies to boot, could render a man immortal? But it seems to be taken as a grievance that a man who had got his life insured died even in spite of that—and Turkish Baths too.—thus making it a matter of blame to them that he did not live for ever. Is not this, however, unreasonable and absurd? He says, further, that he has heard of one or two other cases of persons whose death was referred to the Turkish Bath. But it is all hearsay evidence, as he himself volunteers to say, upon which he ventures to base these rash statements. He fails to give any satisfactory proof or reason whatever for believing that such was really the result of the bath. There was evidently no *post-mortem* examination, and nothing to convince any one that the death of the parties referred to was not, after all, the result of natural or organic disease, which might have happened in ordinary warm baths, of which he speaks highly, rather than of the bath against which the Quixotic editor waxes so unreasonably furious. He knows perfectly well that people have died before now in common baths, which he yet admits to be " powerful instruments in the hands of medical men," and even in bathing in the sea. But it is the fault of the persons themselves in bathing at an improper time, or perhaps when the stomach was overloaded with food.

The grand discovery which the Glasgow editor fancies it has been reserved for him to make is but a mare's nest after all. So far from shortening life, I firmly believe that the bath preserves and prolongs life, and my belief is much more rational than his, being based on a vast amount of evidence, experience, and observation, which it is impossible to gainsay or resist, while his is founded only on the rash and haphazard authority of ignorance and hearsay—a most dangerous course for a scientific, and much more for a medical man, to enter upon. I believe, and I think I have already proved, that the Turkish Bath is one of the most mild and merciful, the most humane and beneficial modes of medical treatment— *without drugs*—ever yet introduced into this country, and that it is the safeguard and preservative of health in those countries where it is regularly used. Why, then, does he so furiously vilify and decry it, but because he perceives all the world is going after it. But let him take warning from the past history of the medical profession, and the humiliation it earned for itself by its long-continued persecution of Jenner, and its fierce opposition to his merciful discovery of vaccination. In casting " firebrands, arrows, and death, he deceiveth his neighbours." And though it may be " sport " to him, it may be fraught with very serious consequences to the interests of truth, in the cause of science, and to the well-being of myriads of human beings, to whom, but for his unjust aspersions, the Turkish or Roman Bath might have proved a saving boon. But I am happy to think that all his brethren are not of his opinion, and that I have every day patients sent to my establishment from the most eminent physicians that this country, and our own city in particular, can boast of.*

Finally, as to the " novelty of the Turkish Bath fast

* As another instance of the violent and absurd tirades sometimes uttered against the Turkish Bath, I may mention that of Dr Lawson of Birmingham, in a small volume recently published, and entitled " A Manual of Popular Physiology," in which there are a great many things that are true, many that are doubtful, and some that are coarse—amazingly so for this year of grace. On the subject in hand he gives no facts—not a single one, but abundance of strong assertions, to every one of which I oppose a total and unhesitating denial. He conjures up heart-disease, liver-disease, kidney-disease, and bowel complaints, " horrible to contemplate," he says, as *possible* consequences likely, or rather in his opinion, reasoning theoretically, certain to ensue from the use of this bath. It is unnecessary to answer a man who argues in this wild and frantic manner, and who shows such profound ignorance of the subject on which he is speaking that it is quite manifest he has never really studied it. He is evidently one of those men who take up prejudices at second hand, and cling to them more firmly than other people do to principles. So ridiculous and absurd are his vaticinations of evil from the Turkish

wearing off," of which the Glasgow editor has no doubt, he is here also quite mistaken. It is, on the contrary, fast spreading. Some medical men may yet be found who, when their patients talk of having recourse to the Turkish Bath, endeavour to frighten them by saying that if they do so they will *melt away*. But patients obstinately persist in being melted away, and with most beneficial results to themselves, for they actually find that the more they are melted the more they gain in flesh, the stronger they become in health, the firmer in muscle, and the sounder in wind and limb.

4. Another popular objection is—*The supposed hazard of subjecting the body for an hour, or an hour and a half, to an atmosphere heated from* 120 *to* 140 *degrees.* This objection is allied to the former, and in replying to it I shall still further answer the one with which I have just been dealing. That the human body is capable of enduring with impunity a very great amount of heat, is a well established and authenticated fact. Thus, we read of Chabert, the "fire-king," as he was called, who withstood the high temperature of a chamber heated from 400 to 500 degrees, while beef was being roasted and eggs boiled, and he was not one whit the worse. This was owing to his strong vital powers. Sir Francis Chantrey's oven, in which his moulds were dried, and which was constantly entered by his men, was heated to 350 degrees. Again, the workmen in the patent slate enamelling establishments in London are reported to

Bath—so foul is his abuse, and so intemperate his language,—that the tirade may be safely left to answer itself. Even a dumb ass, without being inspired by God, might reprove the madness of this prophet.

bear a temperature of from 200 to 300 degrees for six hours daily, not only without inconvenience, but with benefit to their health. Bath attendants, who are frequently labouring longer in the bath daily, are in the enjoyment of first-rate health. In Constantinople the bath attendants live to a great age, some of them up to ninety years of age, in robust health and strength; and the large number of old people in Turkey belonging to the working classes has been generally remarked. It is estimated that four Turkish porters will carry the load of six English; but all, rich and poor alike, take the bath, and are remarkably temperate.*

What is it that enables the human body to bear with impunity high degrees of heat? It is that vital power of resistance to the influence of external physical agents, which enables man to accommodate himself to all the vicissitudes and extremes of climate, and so endows him with a capacity of living in every region of the globe. The fact of which I am now speaking was accidentally discovered in France, in the year 1760, by two French philosophers, who wished to ascertain the temperature of an oven for some special purpose. The female attendant entered it, remained ten minutes there, and marked the thermometer, when it stood at 288 degrees, or 76 degrees above the boiling-point of water. The only effect which this high degree of heat had upon her was to heighten her complexion, but her respiration was not rendered quick nor laborious. This important fact astounded several philosophers. Dr Blagden of Edinburgh repeated the experiment shortly after. On first

* Tucker's Reformed or Oriental Baths.

entering, such highly heated air was very disagreeable, but all uneasiness was removed on the appearance of copious perspiration. But the most important point ascertained was, that while the living body endured with impunity the high temperature of 260 degrees, its own special temperature was only increased from 98 to 102 degrees.

It is to man alone, of all living creatures, that this capacity or power of resisting a very high temperature belongs. In commenting upon some experiments made by Sir Benjamin Brodie, Dr Barter observes that " the lower animals cannot be exposed to anything like such high temperatures with impunity. A rabbit, placed in an oven at not more than 150 degrees, died in a few minutes. And the experiments made by Fahrenheit, as related by Boerhaave, tend to the same conclusion ; for, of various animals shut up in a sugar-baker's stove at 140 degrees, a sparrow died in less than seven minutes, a cat in rather more than a quarter of an hour, and a dog in about twenty-eight minutes. The only explanation of this difference between man and the lower animals in toleration of external heat, is, according to Dr Barter, that their feathers and hair prevent the rapid evolution and consequent evaporation of caloric, by means of which, however, in the case of the featherless biped Man, the normal temperature of the body is in such circumstances constantly maintained.

I have said enough to prove to the satisfaction of every candid mind, that the temperature of the Turkish Bath is not in the least injurious to the health, but, on the contrary, tends powerfully to preserve it, and rapidly to cure and restore it when affected with disease. I have shown that its supposed liability to produce disagreeable effects on the head, and on the body generally, is altogether erroneous. I might quote numerous testimonies to the same effect. For instance, Dr E. Wilson says that the "brain never works more pleasantly than in the thermæ." Dr Rayner says, "I can truthfully affirm that the thermæ produce far less unpleasant sensations in the head, and acceleration of the heart's action, than the lamp-bath, &c. I have found this to be the case, not only upon persons in health, but upon the gouty and rheumatic, and even the consumptive, in whom the pulse is almost always greatly increased in rapidity."

There are circumstances, doubtless, in which it would be imprudent for a person to enter the Turkish Bath, or to stay the full length of time in it during his first or second visit, without the permission of his medical attendant. Where there is a predisposition to disease of the head or of the heart, the bather, even admitting that his medical adviser gives his consent, should not go through the whole process, and should remain for a short time only in the first heated room, and then there is no danger whatever even although such tendency should exist.

But though there is not the slightest danger, there may in some circumstances be disagreeable and unpleasant feelings produced on first entering the Turkish Bath. These, however, are only momentary, and are caused by the sudden transition from the atmosphere without. But the system soon accommodates itself to the greater heat within. The skin becomes softened, and a gentle perspiration begins to break out from every part of the body. This, of course, naturally occurs much sooner in those who habitually attend to the skin, and are in the daily habit of washing themselves. But in the case of the "great unwashed," who permit layer upon layer of dead epidermis to accumulate like a hard horny armour on the surface of their bodies, almost completely blocking up the pores by myriads of minute cast-off scales, and the effete scurf of many dead skins, the disagreeable feeling of tension, and internal heat, will continue until the struggling perspiration at last forces a way out for itself. This, I have no manner of doubt, is the cause of nine-tenths of the inconvenience, or of anything approaching to headaches, ever experienced on a first visit to the Turkish Bath. I do not mean to insinuate that this was the reason in the Glasgow editor's case, but it might well happen with any one. At all events, in this way the slight headache (which soon passes away) may be easily accounted for, without laying it at the door of the bath. And this, too, is what sometimes renders a first, or even a second bath not so immediately successful, nor its benefits so directly perceptible. And yet the benefit to be derived from it in such cases is undoubtedly very great. And generally after the first or second bath, the perspiration comes readily and pleasantly. But for those individuals who are so skin-bound that the temperature of 100 or 120 degrees fails on a first attempt to produce a sensible impression on the secretion of perspiration, whose skins the first attempt has failed to soften and moisten sufficiently to allow the fount of perspiration to open and flow forth freely, it would not be safe to approach a higher temperature at first, or to proceed at once to the Sudatorium. What they should do is—repair to the Piscina, or small cistern of tepid or cold water within their reach, or to the Lavatory again and again if necessary, and by the application of water to the body, cleanse the pores and court their flow. This will, to a certainty, be followed, on returning to the heated chamber, by a rapid outbreak of perspiration, and thus all possibility of headache and determination of blood to the head will infallibly be obviated.

It is the neglect of these simple precautions that has induced some to undervalue the blessings of this bath. Let it, however, receive a fair trial, and be proceeded with under sound advice, and there is no fear but that it will emerge triumphant from under floods of obloquy. But, considering the reckless and indiscriminate manner in which invalids and others have sometimes used the bath, and without any medical advice or superintendence, it is not wonderful that prejudice should in some quarters exist against it.

I would only add further on this point, that the air in the high temperatures of this bath should be free from sensible moisture. If at a temperature of, for instance, 325°, or even much lower, the air were saturated with moisture, instant death would be the result, as no evaporation could take place. Hence it is that common vapour baths cannot be endured at anything like the high temperature which is positively enjoyable and delightful to those C.B.s who take the true Roman

Bath regularly, perhaps weekly. This is one of the particulars, as formerly shown, in which the Roman Bath differs from, and is superior to, the Turkish Bath,—though they are substantially the same. In Turkey the air is rendered moist, steamy, and vaporous in the hottest room. But ordinary air, neither too dry nor too damp, but heated, as in the old Roman Bath, and retaining in suspension just as much moisture as it will naturally draw from neighbouring lavatories, instead of heated humid air and vapour of steam, as in the modern bath in Turkey, is in every respect preferable, as being safer and pleasanter. And where this is the case, and the air, instead of being damp, is moderately dry, and the place well ventilated and other necessary precautions duly attended to, there will not be a vestige of headache nor a possibility of danger.

5. I now proceed to consider another popular objection to the bath—namely, *That admitting the end to be good, it is a violent and unnatural means of attaining that end.* I deny altogether that it is violent; and if the term " unnatural" is used in the sense of artificial, then I admit that it is so without allowing that this is any real objection to its use. Dr Brereton observes, " As to its violence, all I can say is, that I have seen children a few days old go through the process with evident manifestations of delight; and I have seen old men, ready, as they seemed, to drop into the grave, recover, from its use, faculties and vigour which they had lost for years: so far from its being unnatural, it is the only compromise that we can make to a violated nature."

The idea of its being a violent remedy, can only be accounted for, either from the high temperature to which the bather is exposed, or from the supposed violence of the shampooing process to which he is subjected. As to the former, persons unacquainted with chemistry are prone to judge of heat by the figures which mark the temperature in the thermometer, without taking into consideration the *medium* in which that degree of heat is lodged. There is all the difference in the world between *hot air* and hot water of the same temperature. Hot air can be endured by man with ease, with pleasure, and with impunity, at more than double the temperature at which he can endure hot water. While he would be scalded by water at 110 degrees, and by vapour or steam at 120, he can bear for a brief period dry air at a temperature of 500 degrees of Fahr., or even upwards. So that, taking these facts into consideration, it is clearly evident that there is nothing violent whatever in subjecting the body to a bath of hot air of from 120 to 140 degrees, or, as some practised bathers prefer it, of 160 degrees. Instead of its being a violent remedy, then, I consider that it is a principal point in favour of the Turkish Bath, that there is no shock to the system. The whole process is gentle and gradual, and this is owing in great measure to the medium in which the body is immersed, namely, heated Air,—thin air, and not ponderous water. There is a shock in going into an ordinary warm-water bath, or cold-water bath, but there is none whatever in the Turkish Bath. Instead of shudderings and quick catching of the breath, the bather in a well-regulated Turkish Bath experiences nothing but tranquillity, composure, and calmness, both within and without. Thus, I have convincingly shown that the idea of its being violent, as inferred from its nature or its high temperature, is a delusion.

But, on the other hand, some have an idea that it is violent, from the manipulation to which the bather is subjected in the process of shampooing. Now the shampooing process, as described by D. Urquhart, Esq., and previously referred to in the chapter on the Nature of the Bath, may appear anything but pleasant to many of my readers; but let them remember that it is a description of the process when FULLY carried out *in Turkey*, and that it may be modified to any extent, and is in actual practice greatly modified in this country. In fact it may be, and frequently is, altogether omitted, according to the bather's taste, or the physician's prescription. Nay, I am even inclined to think that Mr Urquhart has given an exaggerated account of that process, and that this has had the effect of deterring many from the bath altogether. And my reason for thinking so is, that in a recent communication to the " Irish Agricultural Review," by Archdeacon Goold, relative to a bath taken by him at Cairo, he says :—" The attendant, dipping the camel's hair-glove occasionally in a bowl of warm water, proceeded to operate on my skin. The sensation experienced during the process of friction, was performed in the most *gentle*, yet efficient and artistic manner; the ablution and the soaping were most delicious. . . . . . Then commenced the operation of shampooing, which consisted in gently, as it were, kneading the entire surface of the body. My joints cracked beneath the practised fingers of the operator, but I must say there was no violence used. Neither did I feel any acute, or even sub-acute pain from the manipulation practised." From this, therefore, it clearly appears that anything like violent rubbing, punching, and squeezing of the bather's flesh is altogether unnecessary, and, so far as I am aware, is altogether unknown and unpractised in this country.

Finally, here, the Turkish Bath is accused of being an unnatural means of attaining a good and valuable end. On the contrary, I maintain that it is entirely according to nature—yea, that the principle of it is neither more nor less than an imitation of nature. For instance, when the crisis comes,—or the turn, as it is popularly called, in typhus fever and in some other diseases, the patient falls into a deep sleep, and then a profuse perspiration supervenes, during which deliquescent process all the morbid matters which poisoned the body are carried off. The Turkish Bath imitates this natural process, and enables the body to throw off morbid materials within it, which have either already impaired the health, or, if not removed, will infallibly induce disease. And surely it will therefore be admitted, that it is far wiser, by timely precautions, to bring the body and all its humours into a state of internal purity by the artificial but not unnatural method of the Turkish Bath, than to wait for the access of some violent disease which may indeed carry off these morbid materials out of the body, but is just as likely to carry off the body altogether.

Instead, then, of the bath being contrary to nature,

it is a close and a beneficent imitator of nature. It is Science acting as the handmaid to health. I do not affirm for a moment that the Turkish Bath is superior to Nature's own law for preserving health—namely, *exercise* in order to *perspiration:* " In the *sweat* of thy face shalt thou eat bread." Nor do I maintain that it is a better means than those ordained by nature for promoting a healthy activity of all the functions of life ; which means are, in addition to exercise, pure air and temperance. But when these laws are outraged, violated, and neglected by the sedentary habits, the convivial habits, and the business habits of a nineteenth-century civilisation,—which overworks the brain and underworks the body,—then art steps in, in the shape of a Turkish Bath, and proposes to produce those beneficial effects on the system, and to remedy those evils in it, which nature meant should be accomplished by exercise, air, and temperance. So far, then, from the Turkish Bath being unnatural, it is the lives we lead that are unnatural. I have thus, I trust, entirely exploded the idea that there is anything unnatural in its principle or plan of operation. And I shall conclude this part of the subject in the words of a writer in the " Irish Quarterly Review :" —" The Turkish Bath is an antidote for the unwholesome lives we live,—a peace-offering to outraged nature for our non-compliance with her laws."

6. Another objection sometimes brought against the Roman Bath, and with which I now proceed to deal, is— *That it is a novelty in this country, unfitted for our climate, and unsuited to our busy habits.* This charge also is brought by the learned editor of the " Glasgow

Medical Journal;" but in bringing it, he only succeeds in exposing his own ignorance. Had he been a little more learned, he would never have spoken of it in these terms.

A novelty in Great Britain! Why, it is as old as the first civilisation that obtained in this country. It was introduced nearly two thousand years ago. It is coeval with the introduction into these realms of architecture, coined money, arts, laws, and the fundamental principles of our social civilisation. And it existed in Great Britain, as a necessary adjunct of the civilisation then introduced, even to beyond the middle of the fifth century. Its ruins and remains still exist in the midst of us, as interesting evidences of the presence of that wonderful people, who borrowed it from nations farther east, and carried it with their conquering armies into the nations of the west.

Reared in the midst of the luxury of the Balneæ, it is not to be wondered at that the Romans should have brought with them their longing for the bath wheresoever they went, and wheresoever their victorious hosts forced their way; and that, possessing a mastery over the greater part of this country which they maintained for more than four hundred years, they should have founded baths in their chief settlements here. While Britain continued a part, and a highly cherished part, of the Roman Empire, the bath continued to flourish, and the subsequent barbarism which overthrew and destroyed their great works and buildings, and so long suffered their monumental story to lie buried in the dust, is only to be lamented. But their remains, as I have already largely shown, have been discovered at various places in England, and particularly in Scotland. True, the learned editor of the "Glasgow Medical Journal" did not perhaps know these things, and particularly that grand and flourishing Roman Baths did once exist so near to him in Glasgow as at Duntocher, Croy, Carstairs, and Castlecarry. How irrational, then, is it to call the Roman Bath a novelty in this country, when it is in reality older than anything else that we possess, and when its ruins are the most valuable and interesting that this country can boast of? How absurd to call that a novelty which is really at the least more than 1400 years old!

To declaim against its revival amongst us as if it were a novelty, therefore, is sheer nonsense. Even if it were the novelty which its enemies call it, is it not a very Chinese way of arguing to say, that because a thing is new therefore it is naught and worthless? Why, is not this just the very objection which they and other heathen nations bring against the greatest of all blessings, Christianity itself, when it is first presented to them— namely, that it is a novelty and an innovation upon old customs! But surely such a silly and ridiculous objection is altogether unworthy the philosophic mind of the editor of a "Glasgow Medical Journal."

To say, further, that it is unsuited to our climate, is equally irrational and absurd. After what has been already advanced respecting its existence in Great Britain for upwards of 400 years, at a period in the history of our country too, when, according to all accounts, its climate was vastly more rigorous than it is now—after such a lengthened experience and practice of the bath, it will not do to say now that it is unsuited to this climate. The real truth is, it is independent of climate. If anything, it is more needed in cold than in hot and relaxing climates, where a tropical heat effects in great measure the work of the bath. Accordingly, we find that throughout the vast empire of Russia, and in other hyperborean countries in the north of Europe, hot vapour baths, on the Turkish rather than on the Roman principle, are so common that every peasant's hut has a contrivance for one—namely, by means of stones heated in a furnace, and on which water is thrown. And since it is now the fashion to talk of the Turks as enervated Orientals, &c., because they use the Bath, surely this cannot be said of the Russians, the Laplanders, Finns, and Swedes, or of the ancient Romans particularly, or of the North American Indians, who, it is said, used at one time also something of the same kind. Surely all these cannot be accused of being " sensual Easterns," and " enervated Orientals," " to whose constitutional apathy the bath is well suited." How foolish does the editor of the " Glasgow Medical Journal" look after all this? And, further, as I have already shown, sweating-houses, or just hot-air baths, were till a recent period common in Ireland, and attended with most beneficial effects among the peasantry, who resorted to them. And surely it will be admitted that the climate of Ireland does not materially differ from that of this country.

To suppose then that the bath is unfitted for this climate, because snow falls here, and because this is not a tropical climate, is altogether out of the question.

Why, snow falls in Turkey, as it fell in ancient Rome. Does he think that Rome and Turkey are tropical climates, inhabited only by enervated Easterns? He must go a great deal farther south, and a great deal further east, before he arrives at the mark. Or, if it is to variableness of climate that he refers, has he any reason to suppose that the climate of Turkey is not variable? Travellers assure us that changes of temperature are exceedingly common there. The real truth I believe to be, that the bath is more needed in cold than in hot countries. Some gentlemen here prefer them in cold rather than in hot weather, and take them oftener in winter than in summer. And there is a very good reason for this. In warm weather and in hot countries, the pores are more apt to open of themselves by the copious perspiration thereby induced. Whereas it is in cold countries like ours, and other parts of Europe, that the artificial assistance of the bath is really a necessity, because here the pores are more apt to be closed by our tight-fitting clothes and by the perpetually accumulating quantities of dead epidermis on the surface of the body, and our pores are not so apt to be freely opened by naturally induced perspiration.

As to the statement which he also hazards about the bath being unsuited to our "busy habits," I deem it unworthy of further reply than merely to observe, that when the fault of the age confessedly is that it lives too fast, both intellectually and commercially, making too great a strain upon, and causing too great wear and tear of the nervous energy, we would do well to go back a step to the wisdom of the ancients, and seek occasionally to restore the frame by nature's sweet restorative, REST, in the quietude of body and serenity of mind produced by the Roman Bath.

As bearing on this part of the subject, I here introduce the two following curious extracts, from which it will appear that even the restored and revived Roman Bath is no novelty in this country, but existed in London in 1683, and in Scotland a few years later. The first is from Chambers' "Domestic Annals of Scotland," and refers to the date July, 1702:—" On the principle that minute matters, which denote a progress in improvement, are worthy of notice, it may be allowable to remark at this time an advertisement of Mr George Robertson, apothecary, at Perth. that he had lately set up there a double Hummum, or bath stove, the one for men, and the other for women, approven of by physicians to be of great use for the cure of several diseases. A Hummum is in reality a Turkish or hot-air bath. We find that within 20 years after this time the surgeons[*] in Edinburgh had a bagnio or hot-air bath, and the physicians a cold bath for medical purposes."

The second extract is from a recent number of the *Builder*, from which it will appear that Roman Baths were in actual use in London in the seventeenth century. " Things move in a circle. Saying nothing of the fact that the Romans, when they were here, say sixteen hundred years ago, formed what we now call Turkish Baths, remains of which are constantly brought to light, in the seventeenth century such baths were not uncommon. The description of 'The Duke's Bagnio,' by Samuel

[*] Chirurgeons.

Haworth, M.D., in 1683. would apply to those that are now to be found in all parts of the Metropolis. Dr Haworth says, 'The Duke's Bagnio is erected near the west end of Long Acre, in that spot of ground which is called Salisbury Stables. At the front of it, next the street, is a large commodious house wherein dwells that honourable person Sir W. Jennings—of whose great worth and services to His Majesty at sea, not only the King himself, but almost the whole nation besides, is sufficiently sensible—who, having obtained His Majesty's patent for making of all public bagnios and baths, either for *sweating*, bathing, washing, &c., is the only proprietor of this new building.' On the accession of the Duke of York to the throne, the baths were altered and reopened, under the name of the ' King's Bagnio.' in 1686, by Leonard Counditt, who refers to a bath called the Hummums, near Covent Garden, but says that his establishment is much superior." What will the Glasgow editor say now?

7. I now proceed to notice a number of minor objections, such as, first, *I am too thin*. But it is quite a mistake for people to think that because they are thin they are not fit subjects for the bath. If their condition otherwise requires it, this need be no hindrance. Nay, so far from being an objection, it should be an inducement to resort to it. It is the very thing which they most require. People acquire flesh in the use of the bath. This I have already clearly shown from facts which cannot be controverted. In the bath the entire system of sewerage in the skin is most effectually flushed, and fresh material is introduced. The digestive powers are

increased in vigour, and a much greater amount of nourishing material is absorbed into the general system. Thus, the entire tissues of the body are supplied with a larger amount of solid matter in sound and healthy flesh. It is not at all wonderful, therefore, but only what might naturally be expected, that in the use of the bath a person increases in weight and strength, with abundant elasticity and vigour of mind and body. The Turkish Bath does not and cannot reduce, or bring a person down, unless it is through ignorance carried to excess, or unless it is done purposely by the same means as in the case of horse-jockeys.

Secondly, *I am too fat.* This is undoubtedly a most important consideration, for a fat person cannot, in the nature of things, be a healthy person. And why? Because fat is a disease. The secretion of fat, and its accumulation in the adipose membrane, is almost as rapid in some persons as water in dropsy; so much so, that the older medical writers have termed obesity a dropsy of fat. It is worthy of remark that people afflicted with this disease are generally short-lived. In some cases the accumulation of fat has been enormous. One individual is said to have weighed 52 stone, or 728 pounds; another, $52\frac{11}{4}$, or 739 pounds. In the Philosophical Transactions for the year 1813, it is mentioned that a girl four years old weighed $18\frac{4}{14}$ stone, or 256 pounds. Now, as persons grow stout, and the muscles become interlaced with superfluous fat, the muscular fibre itself is consequently weakened. Many stout persons who may be desirous, and who may feel the necessity of getting rid of this superfluous load—

which is not strength but really weakness—yet dread the usual and obvious remedy—namely, abundant exercise. But the Turkish Bath will effectually retard the progress of corpulency. It will also carry off the existing superabundance of fat, in the shape of copious perspiration. It will consume less time, and be far less fatiguing, and more effectual also, than a violent course of exercise. Over-stout persons lose pounds of useless fat in the bath; and thus, if persisted in, it is a cure for corpulency, and partially equivalent to exercise.

But how, it may be asked, can the same thing be a cure for such contradictory cases as both fat and thin? Why, because in the bath you lose fat and you gain flesh. For, let no one suppose that these two are one and the same thing. A man may be a stout man and yet not a strong man, and I once heard a stately old gentleman declare that he was not so stout as he had been, but that now he was a great deal stronger. Medical authorities have admitted that those who are unhealthily obese lose fat while in the Turkish Bath, but that this is not accompanied with any loss of strength. I confidently assert, then, that these baths are admirably, indeed, I may say, specially, adapted for such cases. They can be applied so as to reduce weight when necessary, but they can also, as I have shown, be used so as to increase flesh. And here I may repeat that several jockeys at the last races in Sydney trained in the bath, and they all bore testimony to the fact that though they dropped weight much faster than by the usual process of gin-drinking and blanket-sweating, they became stronger instead of weaker by bathing, and were firmer in flesh than before.

I shall here present to the reader a case that came under my own observation. A married lady, the mother of several children, had become uncomfortably corpulent and unwieldy. The obesity was increasing so fast as to alarm both herself and her friends. Being on a visit to Edinburgh, and hearing of the baths and of their utility in such complaints, she resolved at once to avail herself of this most excellent means of relief and cure. Her weight was about sixteen stone, with the bath dress only. She was well stewed in the common vapour and lamp bath, and also in the Turkish Bath. The medicated baths were used too, particularly that special one which acts so powerfully on the liver and stomach. The treatment was most energetically and perseveringly carried out for fully five weeks, and with the most gratifying results. The weight was reduced to 12 stone, and notwithstanding the great number of different kinds of baths in such rapid succession, she was not weakened in the least, but felt stronger, and much invigorated, and more elastic than for some years previously. In fact, she had returned to the days of her more youthful years, and was able to walk to a considerable distance without the least feeling of fatigue. In a word, the whole physical frame was vastly improved in symmetry and neatness. Her countenance assumed a very marked and different appearance from what it presented when the treatment commenced. Her pallor and sallow flabbiness were changed into the cheerful and ruddy glow of the most perfect health. No one could look upon her beautiful person without fancying to himself that the immortal Milton must have had his mental eye fixed on such a grand specimen of humanity when he wrote of our first Mother—

> "Grace was in all her steps, heaven in her eye,
> In every gesture dignity and love." *

The lady was quite delighted, as she might well be, with the wonderful transformation effected on her appearance, and returned to the bosom of her family a very different person from what she was when she left home.

---

* Perhaps some might be disposed to consider this comparison invidious, as they are perfectly sure that our venerable mother Eve, being the very perfection of beauty, could not possibly be in any way fat, or nearly so heavy as twelve stone, as this lady still was when she left the baths. But there is a wide difference of opinion among different nations as to what constitutes beauty. In Africa, for instance, beauty consists in corpulence. No matter how fat a woman may be, the fatter the better—even if she cannot walk, if her fingers are like sausages, and she has to be carried in a wheelbarrow, that is the acmé of beauty. They have also a tradition that our first parents were originally black, and that when they took the forbidden fruit, they became pale with guilt, and white with fear. The justly celebrated Dr Guthrie seems inclined to favour this view of the question, as, in a speech which he delivered not long ago in Edinburgh, when introducing a lady of colour who was to lecture against slavery, he remarked that the complexion of our first parents was matter of doubt, and that the African tradition might after all be the right one; and, in supporting the possibility of this opinion, he gave a very beautiful description of the anatomy of the skin, dwelling particularly on that portion of it called the *rete mucosum*, where the colouring pigment is situated, which determines the complexion of the various races of mankind, and humorously remarked that we, the "pale faces" might after all have had the colour bleached out. A certain lady thought this a very impertinent remark; but I leave the Rev. Doctor and the ladies to decide that matter among themselves; or rather I might leave it to the lawyers, whose province it is to make the worse appear the better side, and to prove that black is white, and white black.

This is only one out of many other cases in which a similar happy result has been obtained. " I have recommended the bath," says Dr Rayner, " to several friends and patients, who, although not suffering from any special disease, have still felt ' out of order'—have been getting too fat, have lost their appetite, become pallid and flabby, and suffered more or less from the common result of modern life in large cities, where men exercise the mind too much and the body too little, live in close hot rooms, eat and drink too much, and take a great deal too much medicine. These are just the cases where the bath is likely to be of the greatest advantage." And the following case from the pen of Dr Wilson is still further confirmative on this point :—" My friend B—— is a man of leisure, so far as the common necessities of life are concerned; his worldly career has been successful ; and, in gratitude to the Giver of mercies, he has devoted the remainder of his days to the service of God, to the doing of all the good he can to his fellow-men ; he is largely concerned in the management of public charities of all kinds ; he is regular in his habits, active, and moderate in his diet ; but, in spite of moderation, he is fat, and, as a man who despises personal indulgence, his fat is an annoyance to him, and an incumbrance. ' What can I do to become less bulky ?' said he to me one day. ' Go to the bath,' said I, ' and after the bath walk to your home in Kensington.' ' Impossible,' said he, ' Kensington is three miles away, and I cannot walk the length of a street without panting.' ' Have faith,' said I, ' and do as I tell you.' A week after I received a note from B——. ' I took the bath, as you desired me; after the bath, I felt that I could walk to Kensington, or to Richmond, if I had chosen, but I had an appointment that obliged me to hurry home in a cab. Yesterday I took a second bath: I did walk home to Kensington, no less to my own amazement than that of my family; I ate my dinner with a relish that I had not known for years; and after dinner the power and the desire to walk were so great that I could hardly repress them.' B—— has continued the bath regularly ever since; he looks fresh and well, and more shapely; he knows no fatigue in walking; during the late severe winter he has required no great coat."

But, lastly—*I am in perfect health, and don't require it; it is best to let well alone.* This looks very specious at first sight. But when this statement is narrowly looked into, and brought to the test of actual experience, as the history of mankind in all ages and in all countries has abundantly testified, it will be found that the idea rests on a gross and palpable mistake. And this will appear from the following considerations:—

You are in the enjoyment of the most perfect health to-day: before to-morrow dawns, you may be laid low on the bed of sickness, perhaps of death. Every day's experience presses home upon us all the uncertainty of health and the precariousness of life; and yet it is strange that although there are no truths more patent to every one than these, we are all so little affected by them, and so indifferent to them—as if their very familiarity rendered us the more insensible to the solemn warnings which such dispensations are designed to teach us. And this is the more to be deplored when it is considered, that a very large amount of the cases of disease and death come under the denomination of what are called preventible—that is, that both the disease and its fatal termination might have been prevented had the appropriate remedies been used. It has been calculated by eminent medical men and others, that one hundred thousand persons are at this moment perishing of consumption in these three kingdoms, and as many more preparing to take their place. Whole armies of people are being cut down, and nations are being decimated prematurely. And if so many are cut off by this one disease alone, what an immense number from many other preventible diseases are thus hurried away to a premature grave!

The human frame is a very complex machine. It has been compared to a locomotive engine. The stomach is the furnace, the food is the coals, and the skin is the safety-valve. All know how disastrous and destructive the consequences to a steam-engine are when the safety-valve is obstructed, or is not in proper working order. So, equally injurious, and often equally fatal, are the results of impeded perspiration from the skin. Again, all know that in proportion as such an engine is well kept, oiled and cleaned, and supplied with proper fuel, it will perform well, and answer the purpose for which it was designed for many years. Now, if only half the same attention were paid to the human machine, it does not require the gift of vaticination to say, that it would perform well for 100 or 110 years. In proportion as the functions of the skin and stomach, together with the laws of nature respecting food and air, are attended to, good and sound health will be enjoyed for many years to come. And those, therefore, who say they are in perfect health, and yet neglect the means of preserving that precious boon, are in the unenviable position of him who sleepeth upon the top of a mast.

Do people generally act on this principle in the affairs of common life? Do they not strain every nerve to better their condition in the world? And, if success crowns their efforts, do they sit down contented with what they have already obtained? No, indeed. they do not. The desired point which at one time they thought would be the limit of their ardent aspirations is now as far off as ever; the goal is not yet reached; they must still press on in the race, it may be for gold, or fame, or power; new schemes and enterprises are entered into and eagerly pursued, demanding redoubled energy, and more increased effort than ever. But, in the midst of all this eager chase after the world, two very material and most important considerations are altogether lost sight of—namely, the preservation of their precious health, and the interests which are involved in reference to the spiritual health of the soul, and its prospects for another state of existence, in that Eternity to which we are all fast hasting. I will not presume to enter further into the province of the divine, but the topics of which I am now speaking are so interwoven, the one with the other,—so intimately connected are the healthy condition of soul and body, more so than were ever wedded pair,—that I feel constrained to speak of both. And my object is to show, that the mass of mankind are more earnest and energetic in pursuing the shadowy vanities of life, than in attending to their best interests, both for this world and the next. And although it has been dinned into our ears from our earliest days that cleanliness is next to godliness,—although there is a close connection between the two, and the one is a part of the other; and although the means of attaining both are, by the goodness of our God and Saviour, placed within our reach, yet it is a sad truth, that many are so careless both of their bodily health and of their spiritual well-being, that they neglect their precious opportunities, and are content to drag on a sickly existence of impaired bodily and spiritual health.

Does the mariner act on this principle? Does he sit down and smoke the pipe of ease, when a soft and favourable gale fills his sails, as he pursues his devious path through the trackless ocean, and his gallant ship is careering over unfathomed depths? He knows well that he must be prepared for every contingency that may suddenly arise. He must watch his course and his compass, his charts and his weather-glass, &c., that he may be prepared for the storm. His vessel must always be kept in good trim, and he must be continually on the look-out for squalls. Now, human life is very similar in many particulars to a voyage at sea. It too has its storms and swelling waves, its sunken rocks and fatal strands, its sunshine and its gloomy clouds. Dangers of every kind, seen and unseen, encompass it on every side. But it has also an infallible chart and compass to steer by, so as to ensure its safe arrival and entrance with flowing sheet into the desired haven. The mariner's guides and charts may fail him, but the Christian's are secure and certain.

I shall briefly notice a few of the dangers to which perfectly healthy persons are exposed, notwithstanding their fancied security. It is a well-known fact, that there are many subtle poisons floating in the air we breathe, as also in the water we drink, and sometimes in the food we eat. This has been well illustrated and explained in an admirable work entitled, " Public Health in Relation to Air and Water," by Professor Gairdner, now of Glasgow. In that very excellent and scientific book, the fact of which I am now speaking is plainly and forcibly insisted on, that such morbid and subtle poisons arising from animal and vegetable decomposition, exist around us. Now, if it is true, as already clearly stated, that there are seven millions of pores in our bodies, for the purpose of carrying off, in the form of perspiration, many morbid poisons from the blood, and that, in proportion as this important end is attained, the system is enabled not only to throw off poisons from within, but also to resist more effectually those from without, then what folly is it to neglect the state of the skin, on the plea of letting well alone!

Suppose a person whose skin is in that neglected condition which I have described in previous pages—and there are thousands and tens of thousands who imagine that they are clean when in reality they are not so, whose pores are blocked up with dead matter, and whose blood is thus rendered impure,—suppose, I say, a person to be seized with small-pox, or typhus fever, or some such disease, while the functions of his skin are thus impeded, suspended and inactive,—he would run a very bad chance of his life. Such is, in many instances, I have no manner of doubt, the result of acting on this absurd principle which I have been endeavouring to expose and refute—namely, of letting well alone. In this case, matters are not well, and should not be let alone. Here, as in everything else, prevention is better than cure. And I trust that many who have been acting on the false and mistaken notion which I have now been controverting, will see that it is high time for them to turn over a new leaf, and pay more attention to the functions of their skin than they have hitherto done. If they are at all desirous of enjoying health and long life, let them frequent the Turkish Bath, as better fitted than any other to penetrate the deep cells of the body, which other baths cannot reach. Thus everything noxious will be eliminated from their systems, and their bodies will be rendered almost proof against those subtle poisons which float in the air, which generate so many fatal diseases, and by which so many of our fellow-creatures in all stages of life are prematurely hurried to the grave. For such poisons, finding nothing akin to themselves, nothing which has any affinity to themselves within the purified system of the bather, will be harmless and innocuous in his case, though awfully dangerous to those who have neglected such precautions on the mistaken plea of not requiring them, and of letting well alone.

CHAPTER IX.

GENERAL CONCLUDING OBSERVATIONS.

THE treatment of disease with water application is by no means an invention of modern times. In ancient times this precious element was universally used, not only as a dietetic remedy, but also as a powerful curative agent. It is a remarkable fact, which we would do well to consider, that when pure water was the principal beverage of man, the purifier of his skin, and the chief remedy for all prevalent diseases—so long as his mode of life was thus in accordance with nature, he remained healthy and strong, and attained a high degree of longevity. But whenever this simple mode was exchanged for a more luxurious style of living, debility and diseases of all kinds quickly supervened. But valuable as cold water is in cases where it is suitable, there are many other cases in which it can be of no use whatever, but positively injurious. The same remark applies to every other hydropathic appliance in the abstract, separately and alone; but when we look at it as a great whole, there are very few cases of disease in which one or other of the hydropathic applications will not be found of great service in their treatment.

I had not been many years in practice when my

attention was directed, in a very practical manner, to the wonderful medicinal powers of cold water in certain phases of disease. I was called to visit a farm-servant in the immediate neighbourhood of Edinburgh, who had gone out to his work in the morning in his usual health, and, it being winter, and the cold very intense, with frost and snow on the ground, his left hand became frost-bitten. Every means that his companions could think of was had recourse to, such as rubbing the hand with snow, &c.; but all proved unavailing. I was requested to visit the patient on the evening of the same day, and found the entire hand much swollen and in a state of high inflammation, rapidly extending up the arm. Three of the fingers presented a gangrenous appearance, threatening complete mortification, and severe constitutional symptoms had already set in, with rapid pulse, hot skin—in fact, burning fever and great restlessness. I was much alarmed at the truly formidable nature of the case, and immediately sent for Mr Ferguson, now Professor Ferguson of King's College, London, who even at that early period of his professional career proved himself a good practical surgeon, and a dexterous and skilful operator. He brought another professional gentleman with him, whose name I do not now recollect. After a careful examination of the case, it was decided to cut off the hand at the wrist, as that was considered the only chance of saving the poor man's life. But before proceeding to this very severe operation, it was deemed advisable to consult the celebrated Dr Knox, who had seen much practice in foreign countries, and who was at this period in the very zenith of his fame as a lecturer and anatomist. No time was lost in procuring the Doctor's services; the patient's hand was again carefully examined, and on retiring for a consultation, the Doctor observed, " Now, gentlemen, if you take my advice you will save the patient's hand and perhaps his life, and in the morning you will only require to cut off one finger." The plan suggested was, in the first place to secure the services of a faithful nurse to remain all night with the patient, and to carry out her instructions carefully to the letter, as much would depend on the assiduous application of the remedy; in the second place, the affected hand and arm to be kept stretched on a flat piece of board, and the coldest water which could be obtained immediately applied at short intervals with linen or cotton cloths during the whole night. Dr Knox remarked that no inflammation, however great, could make head against a powerful application like this. On visiting the patient next morning, we were agreeably surprised to find that the treatment had succeeded far beyond our most sanguine expectation. The inflammation and swelling had entirely subsided, only one finger had to be removed, and the patient, although bordering on 70 years of age, was speedily restored to his usual health, and able to resume his accustomed duties on the farm.

This case, as might be expected, left a very deep impression on my mind as to the great value of water, when judiciously applied, as a remedial power of intrinsic efficacy. This impression was still further confirmed by a case which was related to me by that eminent physician, the late Dr Abercrombie. It was that of a young lady, the only daughter of a very wealthy family in London, who had been seized with a malignant fever which prevailed at that time, and was fatal in many cases. The patient had passed through the first stage of fever tolerably well, but the great weakness and complete prostration which supervened, threatened her life. Every possible means which the best medical skill could devise was tried without avail. As a last resource, another physician was called in. He was evidently a thorough-going cold-water practitioner, as will be seen from his curious prescription. He immediately ordered the bed to be prepared in a certain way, and four pails full of cold water to be brought in. The parents were much alarmed, and objected to such formidable means, but he was resolute and determined, as there was no time to be lost. All were therefore put out of the bedroom, except one attendant; the door was locked. The doctor then stood on the bed with one foot on each side of the poor patient, and quickly poured the water over her body in every direction. She was then dried, and quickly removed to another bedroom, and from that hour the convalescence was rapid and permanent, and the young lady was speedily restored to perfect health. The treatment, certainly, was very rough and bold, but the doctor had satisfied himself previously, both from the state of the patient's pulse and skin, that he might confidently anticipate a good reaction; although I verily believe there are very few medical men, even hydropathists, to be found at the present day, who would attempt so hazardous an experiment in any similar case. Yet the result proves that there is a curative power in water far beyond what any one previously unacquainted with its virtues would have dreamed of.

These cases taught me a lesson which I could never forget, of the great value of water as a therapeutic agent in the treatment of many cases of disease, where all other means have proved powerless and ineffectual. During a long course of years I have had many opportunities of testing its merits, in numerous cases with marked success. In my endeavours to carry out the principle I had difficulties to contend with, such as the want of proper means and appliances, the machinery to work with, the frequent opposition of patients themselves, and also that of their friends and the public generally. But I now profess that the desired means and machinery have been discovered in the Roman or Turkish Bath. In my opinion, this is the corner-stone of the hydropathic system. Here we have the proper and effective combination of hot air, hot and cold water, and skilful frictions. Here we have a curative and remedial agent, whose virtues have been tested and proved. It has been shown to be not against nature, but with it, and that it is a most useful and invaluable handmaid to nature, assisting her to preserve the living body in sound health, and to throw off disease in a natural and legitimate way, which is at the same time the most safe, speedy, and inexpensive way. May we not expect, therefore, that the medical profession will candidly look at the facts which have now been set forth, and hail with gratitude the introduction of the Roman Bath as an invaluable boon in the treatment of disease? It is pleasant to see that the *Lancet* declares " It is a powerful agent, of which the virtues are apparent,—an agent of such great power in restoring the active functions of the skin, and the ordinary results of its application are so peculiarly agreeable and invigorating, that it will probably excite the attention of medical practitioners in its relations to disease."

The most common and popular objections against the bath have been already answered; but there are one or two other difficuities against which we have to contend. There is an absence of interest and a want of sympathy with it on the part of the medical profession generally, with, however, some noble exceptions. It is true they do not oppose us openly, except in a few instances; but the opposition is not the less real and formidable on that account. I will refer to one case out of many such, as an illustration. A respectable person, living in one of the provincial towns near Edinburgh, had been ill with severe rheumatism for some months, and under medical treatment all the time; and finding that, notwithstanding all the medicines he had swallowed, he was getting worse instead of better, he asked the doctor if he thought that the Turkish Baths were suitable for his case. The doctor replied authoritatively, that if he wanted to get disease of the heart he might go. The patient immediately answered, "Well, doctor, I have taken your medicines all this time, and am no better, but worse, and am therefore determined to go, and take all the responsibility and risk upon myself." He came, and was cured in two weeks, and, full of gratitude, he returned to his home, and hastened to convince the doctor that his heart was not only quite sound, but that his rheumatism, from which he had suffered so much, had been entirely removed; nor has there been any relapse, although it is some months since the patient returned home. Now, it is very strange that medical men should act in this way. But it is a well-known fact, which has been demonstrated in all ages, that they are generally the last to acknowledge or adopt any improvement.

At a meeting of the Turkish Bath Company in Sydney, in connection with the recent opening of a large bath in that colony, the chairman, Mr Holt, in welcoming those who had assembled at the invitation of the Directors, said, " he desired not to omit to allude to the members of the medical profession who had honoured us with their presence. He was the more pleased to see many medical gentlemen present, because he knew the great antipathy which a large number of the members of that profession had to all improvements. That remark seemed to excite their risible faculties, but he would put it to them if it was not a matter of history that all the great improvements in medical practice had been made in opposition to the Faculty. He had conversed with numerous medical gentlemen about the bath, and he had always found the greatest difficulty, not in persuading them of the advantages of the Turkish Bath, but in getting them even to listen to him while he spoke about it. Therefore, if medical men would shut their eyes and ears to what was going on in the world, what could be expected from them? How could they be expected to learn the value of any new improvement? He was much pleased with a remark that he saw in an address of a professor of medicine to his medical students. that if they wished to maintain their influence over their patients, they must not be behind them in knowledge of the most recent appliances of medical science. He might apply the remark to this colony, by saying, that if medical men would not pay attention to the new discoveries that were being made, their patients would have a greater knowledge of the healing art than they had—so far, at least, as the Turkish Bath was concerned." While these remarks which I have now quoted apply to the profession in general, I beg leave to state most distinctly that there are many honourable exceptions both in town and country,—gentlemen who stand at the very top of their profession, who have themselves introduced many useful improvements in medical practice, and who are always ready to welcome others, when satisfied of the soundness of the principles on which their claims rest.

Another difficulty sometimes met with is the reflex action of these apathetic and hostile views of which I have been speaking upon the public mind, giving rise to a variety of absurd prejudices which are quite erroneous and unfounded. Some people seem to imagine that patients receive rough handling and horrible treatment in the Turkish Bath, and that such is absolutely necessary for its proper administration. For instance, an old gentleman who was exceedingly desirous of testing its merits on his own person, having got his courage screwed up to the proper pitch, as he thought, resolved to make the attempt. But, just when he was on the point of putting his purpose into execution his courage fled, like the pain of toothache on the appearance of the extracting instruments; but, as the needle returns to its pole, his purpose and desire returned with more determined vigour than before. Still, there was yet a certain amount of fear, with considerable hesitation. Seeing this, I encouraged him as well as I could, assuring him that there was no risk or danger whatever. He was persuaded at last to venture in, and afterwards I made out, in the course of conversation, that the real cause of all this perturbation and fear arose from a female friend, who, it seems, was in the habit of whispering all sorts of horrible things concerning the bath in the old gentleman's ear—such as, that if he went to it he was sure to be torn to pieces, or flayed alive, or that he would not come out of it in life. He soon found, however, to use his own words, that all such foolish fears were a perfect bugbear. He was quite delighted with the bath; he left it invigorated and refreshed, and doubtless hastened to convince his lady friend that he was still in the body, and that her ideas were not well founded.

Then there are great objections made on the score of expense. But if the facts already stated are correct, namely, that the Turkish Bath is most powerful as a means of cure, as well as in preserving the frame in a sound and healthy condition, then the small expense necessary to secure these desirable results is a mere driblet in comparison with the large sums of money annually expended on doctors' bills and apothecaries' drugs. Dr Tucker says, "The general practitioners' revenue in England is estimated at five millions annually, while the revenue to the Exchequer on patent quack medicines amounts to one hundred thousand pounds a year. If we compare these most important facts with the personal condition of the barbarians—the Turks—we have some cause for humiliation. In Constantinople there are fewer medical institutions for its population than in any other European city—but there are more baths. In London the medical practice in drugs is very flourishing. There is a very fast trade conducted up and down that grand canal of traffic to general medical practitioners, the intestinal tube. An American medical writer says, 'Medicine is the art of amusing the patient, but Nature cures the disease.' The first Napoleon said, 'Doctor, no physicking; we are made to live. Do not counteract the living principle. Leave it the liberty to defend itself; it will do better than your drugs.' Dr John Armstrong of London, in his "Practice of Physic," stated that it would be well if the young medical practitioners would spend the first year of their practice in drugging themselves; they would then save many of their patients from an untimely end." On the score of expense, then, and in reply to the objection that Turkish baths are dear, I content myself with saying—Yes, they are dear, as compared with other and inferior baths; but they are not dear as a part of your doctors' bills.

Some may suppose that I am Turkish Bath mad. My only regret is that I was not affected with the same mania forty years ago, and my earnest desire is to see the entire community similarly affected. Others impugn my motives as purely mercenary, and ridicule my feeble efforts to indoctrinate the public mind with a knowledge of the great value of Medicated and Turkish Baths in the prevention and cure of disease as altogether selfish—or as a foolish Utopia,—an ephemera destined to flutter in the breeze for a short hour and then die a natural death. And the public press, in many instances, appears to consider the subject as not of sufficient public importance, or worthy of much notice; while, at the same time, if a brutal prize-fight comes off, whole columns are freely devoted to it, and, in the descriptions given of these barbarous exhibitions, the importance of physical culture is lauded, its profound appreciation by the ancients is insisted on, and the training by which these modern athletæ have acquired their gladiatorial and muscular power is commended to the attention of the community generally. But the most important agent in that very process and system of training—namely, the Bath—is coolly ignored and set aside.

Hear, on this point, what is said by the editor of the sporting paper *The Field*:—"Every one who has had to do with bringing either man or horse into perfect condition, knows that the critical part of the process is to apportion the sweats, so that the fat and soft substance may be removed, and yet the stamina of the patient may not be unduly taxed or over-fatigued by the means adopted to produce the perspiration. This result has at last been obtained by the re-introduction to this country of the method which was in practice amongst the athletæ of Greece and Rome, and which, eighteen hundred years ago, was perfectly familiar to the inhabitants of Britain, as the remains of Roman Baths clearly demonstrate. The leading points may be stated as follows :—First, that by the application of hot dry air. the man or horse is sweated, without the incumbrance of heavy extra clothing ; second, that he is thus sweated while at rest, and that therefore the danger attending an over-action of the heart is avoided ; third, that he is sweated whilst naked, and that therefore, by the admission of air to the pores, he is cleansing the system from within, and also engaged in purifying the blood from without. Very recently, Messrs Prince, of the Racket Club, in Hans Place, have opened a bath, realising the Roman process of a gymnasium in conjunction with the bath. The pugilists and pedestrians of the north of England avail themselves of this unfatiguing, yet effective mode of training. The bath at Sheffield is scarcely ever without some of these gentlemen."

As showing also the much wider application of this principle than most people have any conception of, I here give the following extract from the "Scottish Farmer" for 1862. In that volume, at page 1163, it was stated "That Lord Kinnaird had erected a Turkish Bath for the treatment of cattle, and that it had been very effective in the cure of calves afflicted with the scour. It was further mentioned that three beasts, suffering from pleuro in a very advanced stage, had also been subjected to the treatment of the bath at a very high degree of temperature, and that at the time we were there two of them showed manifest signs that the disease had been checked, the symptoms—running at the nose, metallic cough, staring coat, &c.—having been removed, and the animals improving on their feed. The third beast was in a somewhat shaky condition, though not so bad as he had been before he had been placed in the bath. We happened to be at Millhill on Monday; and one of our inquiries there was about the pleuried stots which had undergone the treatment of the Turkish Bath, and from which it was just possible, we thought, the malady might not have been so thoroughly eradicated as not to return again. We were happy to find that the whole three had been completely cured—that they all went on progressing daily after the time of our visit, and had been sold at a price of twenty guineas each. Of course, one experiment is not sufficient to justify the belief, that pleuro will in all cases succumb to the Turkish Bath; but it is sufficient to warrant the trial of this curative agent before the animals are slaughtered. The expense attendant upon the erection of a Turkish Bath suitable for cattle treatment, where there is plenty of water, is a comparative trifle; and as it can be made available for the treatment of horses and sheep, as well as for cattle, suffering under various complaints, it would perhaps be worth the while of large stock-farmers to consider the propriety of erecting such an establishment on their own premises."

In addition to the above interesting facts, I may mention that Mr Wauchope of Niddry, whose beautiful model Roman Bath there, was the first ever erected in this country in modern times, has an apartment attached to it for the treatment of cattle. A truck, into which the cattle are placed, is driven into this adjoining apartment. His is the true principle of the Bath, as regards the proper method of heating. Dr Barter also says, " The bath has been applied by me, and by several farmers in my vicinity, to the lower order of animals with satisfactory results, and particularly in the lung diseases of horned cattle with great success." And Dr Tucker says, " A medical gentleman in the south of Ireland constructed a loose box for horses, heated on the same principle, in which unsound horses are cured by this sweating system. A gentleman at Brackley, in Northamptonshire, adopted the same system with his hunters, and he informed us that nothing could exceed either the enjoyment of the animals, or the rapid improvement of condition and wind, obtained without the necessity of pounding the legs under heavy sweaters."

It only remains for me to add a few out of the numerous testimonies which might be given in favour of the bath. Kind and favourable notices appeared in various public prints on the occasion of the opening of the Sciennes Hill Baths. One newspaper said, " Constructed with singular taste, and at very considerable expense to the proprietor, they afford all the comforts and medical appliances which the most fastidious invalid can desire." Another was kind enough to say, " The establishment of Dr Lawrie, which has now been in active operation for some time, is pleasantly situated at Sciennes Hill, Newington. There are in connection with this establishment a variety of medicated baths, all of which, as well as the Turkish Bath, have been fitted up with great completeness, at a cost, we believe, of at least L.2000. The buildings, with dome and pointed gable, and open spaces dotted with flowers and shrubbery, have an attractive and cheerful aspect. The saloon, bath-rooms, and retiring apartments are models of neatness, everything being of an elegant and soothing character." And, again. "Sciennes Hill Baths are indeed beautiful for situation; they command a fresh breeze from the open country beyond, and a delicious fragrance is wafted from the flower gardens and orchards in midst of which they stand. The public spirit of Dr Lawrie in erecting these baths for the benefit of the citizens deserves a very large amount of acknowledgment, and we are convinced that if the properties of the bath were but fully comprehended, that shall not be wanting at Sciennes Hill." And once more, "The bath-rooms are commodious and elegant, and all the varieties of that healthful luxury — plunge, shower. vapour, douche, and sitz baths—can be obtained at this establishment, which is fitted up in a first-class manner. and in which every provision has been made for the comfort and accommodation of visitors of both sexes. The medicated baths are in communication with galvanic batteries, which are employed to operate upon patients, the infirmity of whose nervous system requires the use of such means for their restoration to health: while there is also an oxygen apparatus with which patients suffering from pulmonary complaints are treated. and in which the healthy action of the lungs is promoted by the inhalation of the oxygenic current. Altogether, the Sciennes Hill Baths are admirable in arrangement, while the secluded situation of the establishment is highly advantageous: and, above all, the experience of Dr Lawrie in treating invalids upon this principle, now introduced for the first time into Edinburgh, affords an excellent guarantee for the success of the establishment." One more brief extract will suffice :—" Indulging the belief, that there was something connected with the Turkish Bath, which more than others exercises a refreshing influence over the wearied frame, we put on the resolution to give it a trial. The process is by no means such as to cause the least alarm; the various stages are but the removal from one delicious situation into another : and at the conclusion of the whole, the bather feels as if new life had entered his system, and he could perform gymnastic feats of a marvellous description. We are glad to know that these excellent baths, which have been erected by Dr Lawrie at an enormous expense, are being appreciated by the public of both sexes, and that the infirm and feeble find in them a panacea for their diseases, and a restorer from the stiff to the agile frame."

The following, entitled " A Visit to the Turkish Bath at Sciennes Hill," and extracted from the " Caledonian Mercury,"—the oldest established newspaper in Scotland, will, I doubt not, be read with interest and pleasure :—" A pressure of mental work, added to our wonted handicraft labour, had unduly taxed our strength, and left us tired, nervous, and miserable : in a word, *used up*. A fortnight at Malvern, a trip to the Trosachs, and a walk over Ben Nevis, were all thought of as likely ' to set us on our legs again ;' and only two trifling matters interposed—time and money. In our misery our eye fell on an advertisement in your journal, ' The Turkish Bath,' and remembering having read of the ' setting up' qualities ascribed to this Eastern restorative, we determined to seek its friendly help. Our way thither lay, not over the golden waters of the Bosphorus, nor through the narrow streets of Baghdad, but down our own delightful Meadow Walk, past the Melville Fountain, and along a few score yards of genuine country road. In the distance, we descried a group of buildings, partly Swiss chalet, partly Turkish mosque, with a preponderance of suburban villa, and a small proportion of British factory; the latter element being contributed by a goodly-sized brick chimney, which, though not a highly æsthetic, is yet a very necessary, adjunct to the establishment.

"Passing up the well-kept garden walk, we enter a pleasant waiting-room, where we are presently joined by the courteous and kindly proprietor of the baths—Dr Lawrie, to whom we state our wishes. Accompanied by our medico, we pass along a passage, in which stands the receipt of custom, presided over by a fair young girl, seated in her little box, with a stained glass-door on either hand. Passing through the inner door, we enter room No. 1, called the *Frigidarium*, where we are met by bathman No. 1, who requests us to take off our boots. This done, our valuables are next demanded from us, and deposited in one of a series of little drawers, the key of which we pocket. We have now time to look about us. We are in a long room, having a window at one end and a door at the other, and gracefully ornamented with statuettes, flower vases, &c. Next the window are a number of haircloth couches; next the door, a series of box-like enclosures, hung in front with green damask curtains. To one of these boxes we are led, and desired to undress, there being left with us a striped calico kilt, and a pair of wooden clogs. We find our undressing-room very comfortable, carpeted like the rest of the divan, and furnished with dressing-table, mirror, combs, brushes, &c.

"Arrayed in our light summer clothing, consisting as above, of striped calico kilt, and pair of wooden clogs, we enter room No. 2, the *Tepidarium*. The hot breath, with which we are met on entering, 'gives us pause;' but, encouraged by a few kindly words from the bathman, we proceed. For a few seconds we feel the atmosphere decidedly oppressive, and are vividly impressed with the idea of a baker's oven. This feeling, however, speedily disappears under the ministrations of the attendant, who, leading us to a tap of cold water in the room, bathes our head and face with the refreshing fluid, which he also allows to run freely on our legs and feet. This operation we enjoy amazingly, and are inclined to prolong; but, receiving a gentle hint that our couch awaits us, we proceed thither and lie down. We are now at leisure, and in a fit state to take a survey of the *Tepidarium*, which we observe is a handsome octagonal apartment, from the centre of which springs a mosque-like dome, the under part pierced with oblong loopholes filled with stained glass, the light from which gives a singularly pleasing and appropriate tone to the room. The centre of the dome forms a highly useful and ornamental ventilator, from which depends a long iron rod, by means of which the bath attendant regulates the admission of fresh air into the apartments. Placed round the apartment are cane-bottomed couches, similar to that on which we recline, having each its pillow and sheet; the use of the latter being merely to protect the body from the rather warm contact of the cane, and not for covering us, as we at first supposed. Further, we notice the tap of which we have such pleasant memories, two ornamental brackets on the wall, bearing decanters, tumblers, sponges, &c. Opposite the door through which we entered is another door, leading to some mystery beyond. We further become aware that we have a companion in the bath, who from a couch opposite gives us welcome. We are soon deep in baths—Greek Loutra, Roman Thermæ, Turkish Hamâms. We discuss the benefits of baths in general, and of the Turkish Bath in particular; and on our hinting that, however beneficial it may be to Mr Twentystone, we fear our spare 'atomy can ill afford the loss of so much of its little fat, we are delighted to hear that our friend Mr Jones increased three stones in weight shortly after having gone through a course of Turkish Baths. By this time we are in a somewhat advanced state of evaporation. Tiny streams of perspiration pour down our breast, and, uniting in a hollow, form a miniature lake. Gradually our talk lags, then ceases. We have lost the sense of our miseries. We lie in a state of luxurious enjoyment, watching the play of the many-coloured light as it flickers on the walls, and floating round the white stucco flowers, touches them into life and beauty. We are of Tennyson's lotos-eaters,

> Propt on beds of amaranth and moly,
> How sweet (while warm airs lull us, blowing lowly),
> With half-dropt eyelids still;

And we cry with them,

> Let us alone. There is no joy but calm!

Our delights are, however, broken in upon somewhat prosaically by the bathman, who occasionally sponges our dripping brow and breast with water, and proffers us liberal libations of the same blessed nectar. The good doctor, too, looks in, now and again, to see us; feels our pulse; asks questions about headaches and giddiness, of which we feel nothing; and informs us 'we are doing well.'

"Having spent what appears an age (a period of time inconveniently indefinite) in this state of calm enjoyment, the bathman at last announces, 'we will do;' on which we rise from our couch of cane, and are led to the mystery that lies beyond, which we are told is named the *Sudatorium*. Here we are given over to the care of an almost naked savage, whose only clothing is a kilt, similar to the one of which he now denudes us. In the centre of the room, the atmosphere of which is 140 degrees, 20 or 30 degrees hotter than that of the room we have left, stands a table covered with a sheet, on which we are placed. Here our savage, knowing that we are perfectly defenceless and quiescent, our sandals taken away, and without the power of removal, does with us very much as he chooses. What he chooses to do is this. We are pinched, and punched, and pulled; we are mauled and knuckled; our flesh is kneaded as if we were so much dough; our joints are cracked; and, finally, we are thrashed as if we were some naughty child or incorrigible truant. All this, however, we take in good part, as it does us no harm; nor does the flaying of our skin, to which we are also subjected, cause us any uneasiness. Our only feeling, as we see the black elongated rolls of effete matter peel off, is one of astonishment as to whence it all comes ; and how it is possible that so much ' stuff' should have adhered to one of our cleanly habits.

"Now comes the washing, which is performed in the *Lavatorium*, a small bath-room off the *Sudatorium*. We are here placed in the centre of the floor, which is pierced with many holes. At a touch of the attendant, a battery opens upon us, deluging us with its delightfully cooling shot. ' Is it cold water ?' we ask. ' Tepid,' is the laconic reply. We are soaped all over, and rubbed carefully with a woollen glove ; we are asked to sit on a water-closet looking apparatus, and have no sooner done so than a masked battery opens on us from below, and another from behind. ' Water, water everywhere ;' and here we sit, feeling new life stirring in every vein. The water is now perfectly cold ; but we are in such a glow of life that we feel that if from some hidden corner another fort should open upon us with frozen mercury, we would neither flinch nor flee. This over, we are dried, enveloped in a sheet, and marched through the *Sudatorium*, where we see another victim stretched on the flaying table ; through the *Tepidarium*, whose couches are all occupied by parties dreaming blessed day-dreams ; into the *Frigidarium*, where, adjusting our classical garment gracefully around us, we extend our beatified body on a couch at an open window, through which the evening breeze, laden with odours from the garden, laps us in delicious repose. As we lie, we feel such a sense of mental and physical power, that we debate with ourselves whether we will arise and solve the Syrian question, or settle the American difficulty, or change our Meadows into a smiling garden, or pull down our dear old Castle's ruins, and make it what it should be, our crowning ornament, instead of what it has too long been, our crowning shame. Which to do first we have not quite settled, when coffee is placed before us, over which we linger lovingly ; at last, however, we get up, return to our dressing-room, and, laying aside our *toga*, resume our ordinary attire, secure our valuables, and, with a hearty leave-taking of the worthy Doctor, take our homeward way—thankful in our heart to Mr Urquhart, who first introduced the Turkish Bath into England, and to Dr Lawrie, to whose enterprise and skill we owe the existence of the first in Edinburgh."

The following kind letter is from Dr Alexander Wood of this city :—" 10 St Colme Street, Edinburgh, 27*th* *September* 1862.—My Dear Sir : I feel bound, at your request, to state that, in my opinion, the baths under your charge at Sciennes Hill are well conducted, and useful in the treatment of various diseases. The Acid Bath has long been used in medical practice, but, from the corrosive nature of the fluid employed, it is objectionable in domestic practice,—and I have availed myself, in many cases of liver and duodenal disease and functional derangements of the system generally, of the excellent baths furnished in your establishment. It is possible the efficacy of the bath may be increased by the current of galvanism made to pass through the body while taking it. Your Barege Baths are invaluable in many rheumatic affections and skin diseases, and, as you know, I have proved their efficacy in numerous cases. Where powerful derivation from the circulating fluid is required, as for the removal of morbific matter, and where it is undesirable to remove this directly, or by the channels of the kidneys or bowels, the Lamp Bath or the Turkish Bath prove powerful curative agents, and as such I frequently employ them. They are agents, however, of such power that they should not be frequently resorted to, and never should be taken except under medical advice. For those who value them for amusement or luxury, it is of great advantage to have your experience in superintending them. In conclusion, it is fair to say that, although I believe you hold what are accounted among us heresies in medicine, I have never heard of your obtruding them on, or insinuating them among, any of the numerous patients whom I have recommended to take your baths.—I am, my dear sir, yours very truly, ALEX. WOOD."

From the House Book at Sciennes Hill I extract the few following complimentary testimonials:—

'I have tried many Turkish Baths in Asia and Europe, and am glad to bear testimony to the superiority of Dr Lawrie's, in respect to the judiciousness of the system, the attention of all in the establishment, and the moderation of the charges.

DONALD GRANT NICOLSON,
of the Inner Temple.'

'Much delighted with the bath, and every thing most satisfactory.

Rev. JOHN DAWSON,
Kirkowan.'

'I consider the arrangements of the Turkish Bath most satisfactory, and the effects delightful.

GEORGE COWAN, M.D.,
Rankeillor Street.'

'I have great pleasure in bearing testimony to the excellence of the arrangements in every department of Dr Lawrie's establishment. Every thing is kept most scrupulously clean, and the attendance on the visitants is admirable. I may add, that for years past, having suffered from biliousness and indigestion, I have satisfactorily proved the efficacy of Dr Lawrie's baths in the alleviation—for the trial has not been sufficiently long for the cure—of those annoying and troublesome complaints.'

Sir JOHN DON WAUCHOPE,
Newton House, Edmonstone.'

'The Master of Torphichen has much pleasure in bearing his testimony to the efficacy of Dr Lawrie's Turkish Baths, in a case of extreme debility after a very severe attack of rheumatic gout.

THE HON. THE MASTER OF TORPHICHEN,[*]
Calder House, Mid-Calder.'

'I have tried this most excellent system of bathing, and can bear most warm testimony in its favour. It is most comfortable and refreshing.

JOHN WILKINSON, M.D.,
Tranent.'

'I consider Dr Lawrie's establishment very complete,

[*] Now Lord Torphichen.

and have no doubt but that the baths will prove very efficacious in many chronic ailments.

CHARLES DYCER, M.D..
Great King Street.'

' I have great pleasure in bearing testimony to the excellence of all the arrangements of these baths. Every thing is kept in the best order, clean, and attendance good.

WILLIAM RIDDELL, C.B.,
Major-General,
of Camieston, The Anchorage, Melrose.'

' After continued fatigue from public speaking, much benefited.

Rev. Dr WEIR,
London.'

' I have taken Turkish Baths in every part of the world where they have been instituted, from Cairo to Sydney ; and I consider that Dr Lawrie's bath is unsurpassed for neatness, comfort, and moderation of charges.

HENRY C. BARNETT,
Physician and Surgeon.
Cosmopolite.'

Finally, as the reader may be interested in learning something further of the progress which the Turkish Bath is making at the Antipodes, I refer again to the inauguration of the new baths at Sydney. A sumptuous *déjéuner* was given on the occasion, after which the chairman called upon those present to come forward and state what the bath had done for them.

The following account of the proceedings is taken from " The Sydney Morning Herald " :—

GENERAL CONCLUDING OBSERVATIONS.   259

"Dr BRERETON thereupon rose amidst considerable cheering, and said :—Mr Chairman, Ladies and Gentlemen, Friends and Companions of the Bath,—It is with no little pride that I rise to respond to the request that has just been made by our worthy chairman. Two years ago, I pledged myself to the friends of the bath in England to use my utmost endeavours to establish it in this colony on a scientific and orderly basis. In the opening of this bath that promise is redeemed. As soon as possible after my arrival in the colony, I converted a house into a temporary bath, in which all that was necessary for merely curative purposes was ensured, although the architectural inaptitude of any building not specially constructed for the purpose rendered it defective so far as comfort and luxury were concerned. By means of that bath a great variety of cases of disease were cured and relieved—many of them of years' standing, and having previously baffled every other method of treatment. Rheumatism, gout, sciatica, neuralgia, diseases of the skin, lungs, heart, liver, stomach, kidneys, and other internal organs, debility, the result of chronic, nervous, and cerebral disorders, drug diseases, and the effects of various mineral poisons—from these maladies, among others, the bath has delivered hundreds, I believe I might say thousands, in Sydney alone. I hold in my hands an interesting proof of two facts,—first, that mercury can exist in, and be eliminated from, the body in a volatile form ; and, second, that the bath promotes that elimination. A patient, who, from his employment, had absorbed an immense amount of mercury, and was so impregnated with it that his symptoms made him as good a prognosticator of every change in the weather as a barometer, became a bather some weeks ago. Since he has bathed, the mercury has passed from his body in such quantities as to cover a gold watch-guard he wore with an amalgam of mercury. I hand round the chain for the inspection of our guests. I will take the opportunity of alluding to two fallacious objections which I find still occasionally put forward. First, that the bath is weakening and unsuited for debilitated patients. In answer to this I have only to state that the weak patients are, I find, those who require to bathe most frequently. Consumptive patients usually gain flesh by bathing. A gentleman suffering from disease in the stomach of long standing, and extremely emaciated, became a bather a short time ago. In two weeks he gained five pounds in weight. It is true that those who are unhealthily obese lose fat while bathing, but this is not accompanied with loss of strength but with increased vigour. The bath may be applied so as to reduce weight when necessary, but it can also, as we have seen, be used so as to increase flesh. Several jockeys at the last races trained in the bath; they all bore testimony to the fact that though they dropt weight much faster than by the usual process of gin-drinking and blanket-sweating, they became stronger instead of weaker by bathing. One of them bathed four times a day, and told me that he lost nine pounds in two days; but that his appetite increased, he had no desire for a drink, and was firmer in flesh than before. The second objection is, that after having been submitted to so great a heat, persons must be very liable to take cold. Now I have had patients far advanced in consumption who could not endure the least change of temperature, and who, after bathing for a short time, have become almost insusceptible of cold. Many people who had been cased in flannels from head to foot for ten, twenty, and forty years, have cast them, not only without inconvenience, but with advantage, and have exposed themselves with impunity to changes which, before they were bathers, would have been dangerous to their lives. I see now before me those who, like myself, would not know how to catch cold were they to attempt it, and who yet once were hardly ever free from it. To whomsoever cold and influenza are terrors, the bath offers a safe, a speedy, and a certain deliverance. In a recent number of the 'Medical Gazette' there appeared a report of a lecture by a distinguished allopathic professor, Spenser Wells, in which he urges the claims of the bath, and points his appeal with the facetious argument that the public now know so much on the subject that it will be prejudicial to the success of medical practitioners any longer to remain in ignorance of it. The principal homœopathic medium—the 'British Quarterly Journal of Homœopathy'—lately contained an article by an old bather, my friend Dr Scriven, of Dublin, in which he urges the importance of the bath to the homœopathic physician as an auxiliary in those cases where medicine alone is insufficient. Within the last two years many of the most distinguished names in medicine and surgery have been added to the ranks of champions of the bath; and baths, more or less perfect, have sprung up in hundreds in various parts of England.

"Mr Mort said, he could not abstain from recording the benefit he had derived from the Turkish Bath while in England. When first he met Dr Brereton (which was in the north of England), he learned from him the value of the Turkish Bath; he was induced to try it, and received great benefit from it. Shortly afterwards he went up to London, and, strange to say, there was not at that time a Turkish Bath in all London. He soon, however, came across a milkman who had met Mr Urquhart, and who had been advised by him to try the Turkish Bath for the cure of epileptic fits. The man had built himself an oven, for it was really nothing else, and to that place he (Mr Mort) resorted, and derived from it such benefits as so incomplete an apparatus would afford. And he claimed the honour of having contributed to the erection of the first Turkish Bath in the city of London. For he had found it impossible to wriggle his body into the place that the milkman had built,—it was nothing more than an oven, and he (Mr Mort) furnished the man with the means of building a better place, where he afterwards regularly bathed. He felt that the public were much indebted to Dr Brereton for his perseverance in this cause, because at the time he first proposed to open the bath in Sydney, no project could have been more thoroughly laughed at. Mr Mort hoped to have the pleasure of erecting the first private bath in New South Wales; his bath would shortly be completed, and he hoped that others would follow his example. It would not be a very costly affair to build a small bath, which might be sufficiently perfect for people at a distance to enjoy the benefits of it.

He thought that people in the country might at a comparatively small expense construct a Turkish Bath, that would at all events be as beneficial as was that of the milkman at which he took his first bath in London.

"Mr P. J. PIGGOTT, from Queensland, on being called upon, said that he rose for the purpose of offering his thankful acknowledgments for the very great advantages he had experienced from the use of the Turkish Baths, which it had been his good fortune to have heard of and to have tested. He might state, to show the benefit he had derived from them, that for many months he had had upon him a severe attack of sciatica, of so painful a character that for a long time he was reduced to the necessity of using crutches, not being able to leave the house—suffering as he did from the disease, not only severe pains in the limbs, but also great bodily debility. Indeed, so severe were the attacks that, although he had the constant benefit of the best medical advice, he continued gradually to get worse. However, he was most fortunately recommended to proceed to Sydney, and to try the Turkish Baths. He need not say that he was only too glad to have the chance of trying any remedy that would afford him relief, although his ignorance of the system adopted at the baths caused him many anxious doubts as to the benefit likely to be derived from them. However, he could now conscientiously affirm that he was a living example of their efficacy, as, after having used them for the last two months, he might truly say, in the language of one of his countryman's songs, ' I'm my own man again,'—while he might with equal truth have said, when he first arrived in Sydney, ' I'm not myself at all.'

" Mr Collie, who was next called upon, had been a severe sufferer from dyspepsia for the last ten years. He had taken a voyage to South America, expecting to get relief there, instead of which his complaint got worse. All the doctors' medicines he took, instead of arresting the complaint, only aggravated it. After he came to Sydney he got worse, until he at length applied to Dr Brereton to know whether he could give him any relief or not, when he (Dr Brereton) advised him to try the Turkish Bath. After he took the first bath he felt scarcely any pain; and after he had taken the third bath, he was perfectly free from pain. A fortnight after he commenced bathing, he found that he had increased in weight seven pounds. His health had much improved, and he hoped shortly to be perfectly well.

" Mr W. F. Jenkins: Although himself a member of the medical profession, he had no hesitation in admitting that the Turkish Bath was regarded by the profession with much ill-feeling. He, however, entertained a very different feeling towards the bath, having derived very great benefit from it. He had been for many years a martyr to rheumatic gout, and had taken a great quantity of physic, which used to give him temporary relief, but he always suffered a relapse. He had taken the Turkish Baths at the time they were commenced in Sydney, and had obtained the greatest relief from them. He must acknowledge having been treated with the greatest respect by Dr Brereton. He might mention, as a proof of what he had suffered, that he had been laid up for seven weeks at a time with gout in his eye. Since he took the baths, however, last Christmas twelvemonths, he had seldom had an attack of the complaint. He considered himself fully twenty-five years younger than he was. As a member of the medical profession, he should consider it his duty to recommend as many patients to the bath as he could; and he had no doubt that before long all other medical men would send their patients here to be cured, not only of the gout, but perhaps most other diseases.

"Sergeant GIBSON next rose, at the request of the chairman, and stated that he had derived very great benefit from the Turkish Bath. He had suffered from rheumatic fever so acutely, that for eleven months he could not raise his hand to his head. After the first six baths he had cast off his flannel, and had not felt the slightest inconvenience; and after bathing for two or three months he had been able to resume his duty as a soldier. For the first two or three baths he was obliged to be carried, but after taking ten or a dozen he was able to walk, and he had been perfectly well ever since. He had lately taken some baths, but more for luxury than for any other reason. He was a perfect skeleton when he first saw Dr Brereton, but now he could walk five or six miles before breakfast.

"Mr HERCULES WATT had been troubled with the gout for upwards of thirty-five years. He had been embedded in flannel from one end to the other, and could scarcely ever go out without catching cold. The baths had done him so much good that he had thrown away the flannels which he had worn for forty years. Now, he couldn't catch cold—try all he could; he had never caught cold since he had taken the Turkish Baths. He thought this institution one of the greatest blessings that could be bestowed upon the country. He meant to stick to the baths as long as he was able. He was nearly seventy years of age, and the baths had made him feel fully twenty years younger.

"Mr DAVY, on being requested to relate his experience of the bath, stated, that in the middle of last October he had tried it for rheumatism, and had found it of the greatest benefit; since then he had tried it for congestion of the lungs; it had restored him to perfect health, and had put him in a better state for enduring fatigue than he had ever been in before. He had on all occasions recommended the bath to his friends, because not only being so valuable in promoting health, it was one of the greatest luxuries that a person could enjoy.

"Mr GARRAN bore testimony to the good effect of the bath in diminishing susceptibility to colds, and in improving the general health. What the Emperor Napoleon called 'the miserable logic of facts' was the most potent sort of argument that could be used, and they had undeniable facts adduced that day. Those who were still sceptical should try a bath for themselves, and that would convince them if testimony would not.

"The Rev. JOHN WEST, in responding to the request to say a few words, observed that, according to the testimony of previous speakers, the Turkish Bath could do almost anything—could make them fat or thin, as they wished. It was also shown that the bath had produced great moral results—it had united in cordial sympathy medical men who were formerly at discord. Besides this, it was alleged that it purified the skin, restored the beauty of the countenance, and gave back thirty or forty years of life,—in fact, that it could do everything for a man in a physical or a moral point of view that he could possibly wish. When we heard of people casting off flannels that they had worn for ten or fifteen years, it was another proof of the great power and efficacy of the bath. He remembered one day a man telling him of his having left off the drawers that he had worn for fifteen years. He had seen enough of the Turkish Bath to be convinced that it was invaluable as a curative process; and although he could not present himself as a pre-eminent specimen of the power of the bath in reducing flesh, nor as having absolutely made him twenty years younger—nor even as having restored great beauty to his countenance—nevertheless he could say that having suffered from abstinence and hard labour, and from occasional fits of gout, and having tried the Turkish Bath as a curative process, although he had had one severe fit after he took the baths, he had not had another for some time.

"The Chairman then concluded the proceedings by inviting the gentlemen present to partake of a bath. The company having retired from the breakfast table, about twenty gentlemen availed themselves of the invitation.

"If it be true, as has been contended by many eminent

writers, that the physical superiority of the ancients was in a great degree owing to their regular and systematic use of the bath, and that by a peculiar combination of air and water· they were enabled to retain throughout their days undiminished health and vigour, it must be matter of gratification to the citizens of Sydney to know that they have now at their command those advantages which the Romans and other nations of antiquity turned to such good account. In the erection of the new Turkish Bath, the inauguration of which was celebrated yesterday, an institution has been established in this city similar in its main features to those luxurious and costly baths in which it was the wont of the ancient Roman emperors and patricians to indulge, and the ruins of which are still to be seen. The efficacy of the bath as a curative process has been abundantly and satisfactorily demonstrated—numbers of persons willingly and gratefully testifying to the relief it has afforded them from protracted and disabling ailments. The inauguration of an institution which has already performed such remarkable cures, and which offers to repeat those cures for the benefit of thousands now suffering, is an event of peculiar significance and interest in the history of this community."

The Directors' report of the Turkish Baths in Melbourne illustrates cases of equal importance.

"Extract from Mr Rogers' Speech, upon the occasion of his Benefit at the Theatre Royal, Melbourne, July 26, apologising for the postponement of his Benefit, owing to his severe illness, from which he had been restored in a way that impelled him to make known its efficacy to others.

"The fact is," said Mr Rogers, "I've had a Turkish Bath; I crawled into it, and jumped out of it,—not because I was hot, but because I was cured."

CHAPTER X.

BAREGE AND ELECTRO-GALVANIC BATHS, ETC.

These baths have now been in full operation for more than four years, and have been highly appreciated by the public, and eminently successful as a means of cure. They have been attended by upwards of two thousand patients afflicted with various formidable diseases, which had for years resisted all forms of medical treatment, the great majority of which, however, have readily yielded to the influence of the Medicated, Galvanic, and Turkish Baths. As these baths, mentioned above, have been but recently introduced into Great Britain, and are the only complete baths of the kind yet established in Scotland, a few words in reference to their history, first of all, will not be out of place.

The Barege Bath has been justly celebrated from remote antiquity for its salutary effects in rheumatism, gout, and skin disease, from the slightest to the most inveterate, and in a great variety of the various ills which afflict humanity. It is so called from a place of the same name—Barege, in France—famed for its medicinal springs. Julius Cæsar, and the Roman general Sertorius, bathed in the waters of Barege, to restore their wonted energy after their campaigns in Gaul and

Spain. Henry IV. of France frequented them in his youth; and Louis XVI. dignified them with an hospital for his wounded officers, and another for his soldiers, who, when past all other means of cure, were brought to the baths from the remotest parts of France, as a last and sure resource. The high reputation which they had acquired continued to be sustained in more modern times, and at length led to their imitation and adoption at other places than their source.

During the revolutionary wars in France the number of disabled and diseased soldiers became so great, the crowd of invalids in the hospitals at Barege being at the same time so immense, that it was found impossible to send them all to the natural thermal springs at that place, as had formerly been the custom. In these circumstances the Government ordered the most eminent chemists of France to analyze as carefully as possible, and discover the chemical composition of the waters at their source. This was done, and after the scientific analysis had been made, and the chemical contents of the waters had been discovered, it was found possible, with the knowledge of their nature and properties thus acquired, to make artificial baths of exactly the same chemical qualities. Various hospitals in different parts of the country were then ordered to be provided with the artificially medicated baths. This artificial Barege Bath was much used and highly prized by the first Napoleon. It has been satisfactorily proved that the natural waters of Barege, Aix-la-Chapelle, and Bourbonne-les-Bains, can be exactly imitated by art. The formula for their composition, with the varied and numerous ingredients in their necessary proportions, was adopted and authorised by the Imperial Government. And it was ascertained by the published statistical accounts that the artificial effected as many cures as the original Barege waters at their source. Thus chemistry has happily put it in the power of man to imitate with scrupulous exactness all mineral waters, and to accommodate the heat not only to the state of the patient, but even to vary it according to the varying scale of these natural hot springs. And thus the benefits to be derived from them may be obtained by invalids and valetudinarians at home, without the fatigue and expense of travelling to distant lands or foreign watering places. I may state here, that the artificial Barege Bath used at this establishment is carefully prepared from the prescription before-mentioned, as sanctioned by the most eminent French physicians, and authorised by the Government of that country. Various persons who have had long experience of these baths at their source, have uniformly declared that they could not detect any difference in the artificial from the natural waters, either in the smell, colour, or efficacy of the bath.

Several works descriptive of these baths have been published. Sir Arthur Clarke, M.D., affirms that more cures are recorded by the Barege waters than by any other medicinal spring in Europe. M. Dessault published an essay, recommending their use in stone and gravel. And Sir C. Meighan advocated their use in the cure of gunshot and other wounds; also in muscular contractions, scirrhous tumours, and many other disorders. But their general use is for disorders of the skin, gouty, rheumatic, rigid, and palsied limbs, and cases of painful wounds. Sir A. Clarke was the first to introduce this bath into Ireland, in 1820 ; and he relates some very interesting cases of individuals who were cured by it in an incredibly short period, and where other remedies, after a long trial, had proved unavailing.

Without entering further into detail, I humbly submit, that, during thirty-six years' medical practice in this city, I have never met with any therapeutic agent so efficient, and so prompt in its restorative action, as witnessed in those invalids who have availed themselves of the use of this bath, especially when combined with the Turkish Bath ; and having been in the habit of prescribing its use for more than six years, I can bear ample testimony to its great value and worth, as an admirable remedy for many cases and kinds of disease which no other known therapeutic agent can reach.

The next bath, which justly claims equal prominence, is the Nitro-Muriatic Acid Bath, invented by the late Dr Scott of Calcutta, and successfully used by him for many years, as a remedy for many morbid conditions of the liver. The late Duke of Wellington, when in India, laboured under a severe liver affection, and was restored to health in a short time by the use of this bath. Dr Scott affirms that he employed this bath with decided advantage in almost all cases of morbid secretion of bile, whether the secretion were morbid or depraved ; and that in the paroxysms of pain from a gall stone passing the bile-ducts. or from spasm, he found it to act like a charm, and produce almost immediate ease. The doctor was of opinion, from the rapidity with which it acts in some cases, that it operates on the nerves and not by the absorbents, and he infers, from various experiments, that it is the chlorine of the muriatic acid alone that is the beneficial ingredient of the bath. But subsequent experiments have proved that this bath admits of a far wider range, and is destined to occupy a position as an important therapeutic agent, far beyond what the original inventor had any conception of.

Its detergent properties were accidentally discovered in 1852, at New York. A man, occupied with electroplating, had immersed his hands into solutions of nitrate and cyanure of gold and silver, whereby a dangerous ulcer was caused, which resisted the most energetical remedies; at last the patient plunged his hand into the Electro-Chemical Bath at the positive pole, and after a quarter of an hour the metal plate connected with the negative pole was covered by a thin layer of gold and silver. A few more applications of the Electro-Chemical Bath proved sufficient for the cure of the ulcer. Hence the discovery and introduction of this bath for the extraction of metals and other deleterious substances from the body. Dr Caplin of London, and Mr Hardy of Harrogate, were, I believe, among the first to introduce it into this country, and both of these gentlemen have been very successful in their treatment of many distressing cases of disease arising from metallic poisons in the system.

In a work by Dr Althaus of London, it is stated

(p. 324), that we are indebted to M. Poey, an eminent meteorologist, for the discovery of the Electro-Chemical Bath, and that this gentleman has proved by a number of experiments that it is quite possible to extract metallic substances out of the human body by the aid of electricity, whether such substances have been taken as medicines, or have been lodged in the body by absorption, in the different arts and trades in which their employment is required.

According to Dr Althaus, the Electro-Chemical Bath was at first administered in the following way: The patient is placed up to the neck in a large metallic bathing tub, which is filled with water, and insulated from the ground; he sits in the tub upon a bench of wood, insulated from the tub, and about the length of the body. If mercury, silver, or gold is to be extracted, the water with which the tub is filled is acidulated with nitric or hydrochloric acid; if lead is to be extracted, sulphuric acid is added to the water. One extremity of the tub or bath is connected with the negative pole of a galvanic battery by means of a screw; and the positive pole is held by the patient alternately in the right and left hand. The galvanic current now enters the body, and circulates, according to M. Poey's graphic description, from the head to the feet, traverses all the internal organs, and even the bones, and seizes every particle of metal which may exist anywhere, restores it to its primitive form, and deposes it on the whole surface of the sides of the tub from the neck to the feet, and always more abundantly over against that part of the body where the metal is supposed to exist.

Dr Althaus objects to the latter part of this statement, and is of opinion that there must be some mistake, as he cannot understand how the galvanic current can convey into the liquid of the bath, and diffuse on the whole surface of the sides of the tub, metallic atoms, which, according to the established laws of electro-chemistry, ought to be deposed only upon the surface of the electrodes. Many other eminent physicians are of the same opinion, and do not credit the astounding assertions of some parties as to large quantities of pure metallic mercury being extracted from the bodies of patients whose systems have been charged with that dangerous drug.

Dr Caplin of London, who claims to be the first that introduced the bath into this country, says, in page 97, in reference to this point, that some parties in America assert that the mercury is found in a pure metallic state in the bath, and endeavours to prove very satisfactorily that such is not correct. The doctor says, it is absurd to suppose that the effect of chemical-electricity should be to pick out the metallic molecules, without acting at the same time on the substances contained in the matter of perspiration, so fully described by Erasmus Wilson in his valuable work on Diseases of the Skin.

But it is not my intention, on the present occasion, to enter further into the discussion of this point. So far as my own experience goes, although I have not been able to detect quantities of pure mercury in the liquid of the bath, yet I have had abundant proof of the powerful detergent properties of this bath, in the cases of those who have been dosed with mercury. It is true, that although we have at our establishment a copper bath, we prefer the plan of Mr Hardy of Harrogate, and at Sciennes Hill we use instead the enamelled porcelain baths, which are equally efficacious and more cleanly. Besides, the copper bath is objectionable on account of the immense labour required in keeping it clean, and also from the action of the acid upon the metal itself, thereby rendering it very difficult to determine the real nature of the ingredients deterged from the body of the patient.

The Electro-Chemical Bath has been found of immense benefit in a great variety of distressing maladies, which had for years resisted every other mode of treatment, however skilfully or energetically applied. Without, however, entering into a minute detail of the various ailments successfully treated by the electric current, I will only observe, that it has been proved that there is a striking analogy between the nature of the nervous fluid and electricity. And it has been clearly defined, that the electricity existing in the body may either be in too great abundance, locally or generally, or there may, on the other hand, be a great deficiency of this vital agent in the system. Each of these abnormal conditions is the prolific source of a variety of diseased complications, such as general debility, complete prostration of nervous energy, diseases of the liver and stomach, rheumatism, rheumatic gout, sciatica, and a host of others too numerous to mention. Now, it is a distinguishing feature of electricity, that it can be made to meet the necessity of either case, by supplying the system and setting free the polarized electricity therein, or in diminishing the quantity already existing. And this important vital energy is thrown into the system, or exhaled from it, without a shock and without pain, while the patient is immersed in the bath.

Thus it is that the Electro-Chemical Bath presents to the observer important results, by propagating through the particles of the human tissues, with more or less energy, the dynamic force generated in the pile. And, as bearing on this subject, in the " Philosophical Transactions" for 1807, a remarkable observation is recorded by Sir Humphry Davy, to the effect that, when he immersed his fingers in a vessel filled with distilled water, and connected with the negative pole of a voltaic pile, alkalies were extracted from the body and deposited in the water ; but if the vessel was connected with the positive pole of the pile, phosphoric, sulphuric, and hydrochloric acid were deposited. Wollaston and Cruickshank have shown the same thing ; and hence it has been concluded that the secretive action of our most important organs, as the skin, stomach, liver, intestinal canal, &c., is really the result of the electrical condition of these organs. And, as in all cases of secretion, digestion, sensation, and motion, the nervous influence, which appears to be almost identical with electricity, is the great natural agent, we can thus appreciate the importance of judiciously administered electricity of low tension, in restoring a healthful condition to the secretive and digestive organs. For, since health depends on the equal distribution of this fluid, it follows that when it is lost or suppressed, or in any respect injured, the stimulus imparted to the secretions is proportionally withdrawn, and they become incapable of performing their offices, by which excrementitious materials are retained in the system, to the general detriment of the health. It becomes, therefore, an important object to restore these suppressed secretions by the application of heat or vitality, which is synonymous with electricity. And all diseases, of whatever character or name, being the result of morbific matter and obnoxious substances of some sort retained in the organism, the cure depends on the extraction of these obnoxious substances; and it is confidently believed, that there are no means of eliminating them from the system equal to the Electro-Chemical Bath.

In conclusion, it may be observed that there is nothing alarming even to persons of the weakest nerves, in the application of the Electro-Galvanic Bath. People generally connect the idea of a sudden shock with electricity or galvanism. But those who have tried these baths have been quickly disabused of this mistaken notion; and so far from there being anything disagreeable in them, many have confessed that the new sensation afforded them in the baths is one of the most pleasurable and delightful they ever experienced. A gentle, and yet constant, current of vital energy seems being infused into the system, languor disappears, and a new feeling of lightness and vigour pervades the frame. So that, even apart from the known and proved beneficial influence of these baths as a most valuable therapeutic agent, it might be said that the discovery of a new sensation and a new pleasure, for which a Roman emperor proclaimed a reward, has now been made. So far, therefore, from there being anything in these baths to deter any one from partaking of them, they present a most agreeable, encouraging, and merciful relief to invalids generally; and it may be affirmed, with the utmost confidence, that they only require to be more widely known to become highly appreciated and extensively used.

I now proceed to adduce several cases, out of the many which might be brought forward, illustrative of the benefits of these Medicated Baths, and of the electro-chemical treatment :—

No. I.—Case of Severe Rheumatism of long standing.

10*th April* 1860.—Mrs M., aged 46, has been afflicted with rheumatism over the whole body for three years and nine months, and during that time was quite unable for any kind of work, or even to put on her own clothes. She was also troubled with swellings, or what is commonly called Blind Boils, under the skin of both legs. Various remedies were tried without benefit. The medical attendant said that no medical treatment would be of service, and that she must trust to the efforts of her own constitution ultimately to throw off the disease. Such was the condition of Mrs M. when she consulted me. I prescribed some remedies and ordered hydropathic applications at home. These were persevered in for some time, with very slight alleviation of some of the more pressing symptoms; and on the 10th April, by my advice, Mrs M. commenced the baths at Sciennes Hill, and attended regularly for three weeks, frequently twice in the day, with the most happy results, even a complete and perfect cure. Previous to her illness, Mrs M. had been in the habit of filling up her leisure time in her family by dress-making, and in this way aided her husband in the support of the family. During the whole of her long and severe trouble she was unable even to hold a needle in her hand, but now she is again actively employed, and, since the beginning of May, she has made with her own hands no fewer than thirteen ladies' dresses.

No. II.—Case of Skin Disease (which will be best detailed in the words of the individual himself).

EDINBURGH, *Sept.* 1860.

DEAR SIR,—I feel great pleasure in complying with your request, and in certifying the almost immediate cure I derived from taking the benefit of your Medicated Baths. Last autumn I suffered greatly from a remarkable itching irritation over various parts of my body, in spots about the size of a shilling, chiefly around the neck and arms and under parts of the body, and when going to bed it was almost like to drive one distracted. I tried every kind of medicine, both internally and externally, but to no effect. After a course of your medicated baths I was completely cured, and up to this time, twelve months since, I have happily had no return. Those afflicted in a similar manner I would advise by all means to try your baths.—I am, dear sir, yours very sincerely, T. I.

No. III.—*Case of Acute Sciatica.*

Mr ——, a young man about twenty-five years of age, had suffered severely for six months from the above complaint, had been under medical treatment for the most of that time, and had become quite unable to attend to his ordinary duties as a clerk in a provincial town. All kinds of remedies were tried, and even the needles were inserted into the affected part, without the slightest benefit. The health rapidly declined, the patient was quite emaciated, his complexion was very pale, and he had the appearance of one in the last stage of consumption. His appetite was entirely gone, and his spirits were sunk and flat. Mr —— was brought to Sciennes Hill, 29th January, in a cab, and required the assistance of two men to enable him to come into the baths. The pulse was small and very quick, with a dry hot skin. I entertained a very unfavourable opinion of the case, and had grave doubts as to its final issue. After the fourth bath, however, the patient began to improve. There was less pain in the joint, he could walk a little without assistance, the countenance was cheerful and animated, he felt a greater desire for food, and altogether the improvement was very marked and rapid. The baths were continued regularly until the 22d February, when Mr —— found himself able to resume his ordinary employment. I heard of him from his father in the month of August, and he was then perfectly well, and had had no return of the complaint.

No. IV.—*Case of Rheumatism.*

135 Causewayside, Edinburgh,
24th September 1860.

Esteemed Sir,—Deeply convinced of the benefit of which the public would be deprived, were I to refrain from making known the speedy and effectual relief which, under your skilful and ingeniously devised treatment, and during a fortnight's application of your medicated baths, I obtained from rheumatism, when all ordinary medical, and even Infirmary treatment, had alike failed to perform a cure, I have great pleasure in embracing the opportunity afforded me of bearing my unqualified testimony to their efficacy.

Allan Cameron.

To Dr Lawrie, Sciennes Hill.

No. V.—*Case of Eczematous Eruption on Face and Hands.*

Mrs D——, aged forty-seven, of a florid complexion and sanguine temperament, and of a strumous habit of body, had been troubled for fourteen years with the above complaint, which resisted all known remedies both external and internal. The eruption was characterised at times by intense and intolerable itching, and a watery exudation of a corroding nature, irritating the sound skin in the immediate neighbourhood of the affected parts. She was perfectly cured in about twelve weeks, with the medicated, vapour, and barege baths, and has remained perfectly free from any return of the complaint up to this date. I might enumerate many similar cases of this very obstinate and troublesome complaint completely cured; while there were others with whom we were not equally successful, from the long standing and inveterate character of the disease, and partly from a want of persevering patience in the patients themselves where there was a reasonable prospect of a favourable result. This remark applies also to a variety of other cases of disease.

No. VI.—*Case of Functional Derangement of the Stomach and Liver.*

In addition to those cases already mentioned, there have been not a few persons sent by eminent physicians in Edinburgh, chiefly cases of indigestion and sluggish action of the liver and stomach, all of whom have expressed themselves as greatly benefited by the treatment at these baths. I will only mention one of these, a sea-captain, who had seen much service, both in hot and cold climates, and who had suffered for a considerable time from functional derangement of the liver and stomach. The appetite was various and indifferent, and he had little relish for food, which caused him much suffering from the slow and imperfect digestion. His medical attendant had tried a variety of remedies without benefit, and, having much faith in the Nitro-Muriatic Bath, recommended the patient to try a course of baths, which was perseveringly carried out with prompt and marked improvement in all the symptoms. The captain repeatedly expressed himself as having derived more benefit from them than from anything else he had ever tried, and regretted exceedingly that engagements in London, in connection with his duties as captain, prevented him from making a longer stay, that he might enjoy—to use his own words—" the luxury of these baths, which had already done him so much good." Only eleven baths were taken.

No. VII.—Severe Case of Strumous Ophthalmia.

Miss S., aged twenty-one years, had been suffering for three months from the above troublesome complaint, had during that time constant medical attendance, and all the remedies and appliances consistent with what is called active practice were systematically and energetically carried out. The friends of the patient becoming alarmed, and having a firm conviction that the medicines used were evidently inflicting a more serious injury upon her constitution than the disease they were intended to cure, resolved to abandon all medicines, and to try a course of hydropathic treatment. To this they were advised by their minister, who was deeply interested in the case, and can, if need be, testify to the truth of what is now stated.

Miss S. came to Sciennes Hill Baths on the 13th of August last, and at that time presented the following appearance:—Body much emaciated; face pale, puffy, and bloated; tongue dry and red; breath fetid; gums inflamed and spongy: constant pain in the eye, extending to the back part of the head; great intolerance of light, and running of hot tears, irritating the cheek on the affected side; the eyelids inflamed and swollen, and very little sleep at night: appetite almost gone; the pulse quick and wiry, and the skin hot and dry. The treatment was immediately commenced, and perseveringly continued until the 1st of October, when the patient was happily restored to a more perfect state of health than she had experienced for a long time previous to her illness. The improvement was so rapid and marked, that her friends and acquaintances, who had seen her when labouring under the disease, hardly knew her after her recovery.

This patient's system had been charged with mercury to a greater extent than any that we have yet met with since the Baths were opened. And, as a proof of their successful action, it may be mentioned, that the effete matters eliminated from her system through the pores of the skin, while in the bath, were so offensive, that the bath-maid declared, that they excited within her an insupportable feeling of sickness and nausea: and that, although she has been a bath-attendant for more than twelve years in various hydropathic establishments, she had met with only one similar case in all her former experience. It was extremely pleasant, however, to observe the very marked and daily increasing improvement in the patient's health, in proportion as the poisonous matters were deterged from the system; and thus, in the short space of six weeks, this truly distressing malady was completely removed. Such a case as this, presenting as it does so strong and decided a contrast between the perfectly restored health of the patient and her previously shattered and miserable state, affords a most complete and satisfactory demonstration of the benefits resulting from these baths.

The following certificate is from the minister above referred to:—

17 Greenhill Gardens, Edin,
17th October 1860.

My Dear Friend,—I am very glad indeed to bear my testimony to the truly wonderful cure effected in the case of Miss S., an account of which you have enclosed for my perusal. I have also had the great happiness of observing similar effects produced by your admirable treatment in several other cases. Most heartily do I wish that many, many more of the afflicted had the benefit of your system.—I am, most truly yours,

John Kirk.

To Dr Lawrie.

No. VIII.—Case of General Debility and Rheumatism in Lower Extremities.

Mrs ——, a young married lady from England, had been recently confined with her first child. Her recovery, however, was not satisfactory. One of the breasts inflamed and suppurated, and the general health was much affected. Severe rheumatism attacked the lower extremities, and caused much suffering, especially at night, depriving her of sleep. Having heard of Sciennes Hill Baths, she resolved to avail herself of that means of cure, as soon as it was possible to be removed to

Edinburgh. Mrs —— arrived in the beginning of September, when I was immediately sent for, and found the patient dressed and lying on a sofa. She complained very much of pains in all the joints of the lower extremities, feet and toes, and could not be moved in any direction without the most excruciating agony. Her pulse was very feeble and rapid, countenance wan and anxious-looking, having a cadaverous appearance; the whole body, especially the limbs, very much attenuated; the skin hot, dry, and shrivelled; and the tongue scruffy, white, and furred. Mrs —— was brought to Sciennes Hill Baths in a perambulator on the 10th September. The Alkaline-Galvanic Bath, being the one indicated from the nature of her disease, was taken daily. After eight baths had been taken, the patient was so much improved that she was able to dispense with the perambulator altogether, and walked to and from the Baths on her own feet. Soon she was seen walking in various parts of the town, perfectly restored to sound and vigorous health, and at the end of two weeks was able to return home to her friends, who were truly astonished to witness so great a change in so short a time.

Her uncle, who had some business to transact in Edinburgh, could not leave until he had availed himself of the opportunity of taking a short course of the same kind of baths which had had such a wonderful efficacy in the case of his niece, and expressed himself as quite delighted with the luxury of the Alkaline Galvanic Bath; intimating also his intention of returning in the spring to enjoy a longer course of these valuable baths.

No. IX.—*Case of Derangement of Stomach, accompanied with Total Deafness.*

EDIN., 15*th October* 1860.

DEAR SIR,—I have much pleasure in bearing my humble testimony to the efficacy of your baths, as a powerful remedy in affections of the stomach and liver. Having suffered for some time from indigestion and general debility, with almost complete deafness, I was, through the use of your baths, restored to perfect health, and my hearing is now as good as ever it was in my life.

I am happy also to state that my wife, who had been in very indifferent health for a considerable number of years, from a chronic disease of the liver, which did not yield to the best medical treatment that could be procured in Edinburgh, after having gone through a course of your invaluable Galvanic Medicated Baths, was completely restored to good health; and up to this time, nearly six months, there has been no return of the complaint.

Trusting that others similarly affected will lose no time in availing themselves of a method of cure so pleasant and so effectual,—I remain, yours faithfully,

D. C.

To Dr LAWRIE, Rankeillor Street.

Many other interesting cases might be adduced, but, as they would swell this volume to a much greater length than was originally intended, I shall now proceed to notice the

INHALATION OF OXYGEN GAS.

This gas, as is well known, forms a most important and indispensable element in the composition of atmospheric air—that is, the air we breathe, and by which combustion, together with animal and vegetable life, are sustained. Atmospheric air is composed of so many parts of oxygen and nitrogen, so nicely balanced that the one keeps the other in check; for, if there were a superabundance either of the one or the other, the result would be the destruction of all animal and vegetable existence. According to Mr Samuel Parkes, of London, oxygen gas in its pure state has the property of accelerating the circulation of all the animal fluids, occasions the most rapid combustion of all combustible substances, and is the most powerful and energetic agent with which we are acquainted.

But my object in these remarks is briefly to consider oxygen as a therapeutic agent in the cure of disease. It was used by a number of physicians as a remedy about the end of the eighteenth and the beginning of the nineteenth century. An interesting account of the use of oxygen gas as a remedy in various intractable diseases, with a detail of the cases successfully treated by it, will be found in a small volume published in 1857, by Dr Birch of London. The Doctor says, in page 3, " There is nothing new in suggesting oxygen as a remedy. My professional brethren must needs all be aware that, towards the close of the last, and beginning of the present century, it was used by Drs Beddoes, Hill, Thornton, and several other physicians and practitioners of medicine, in numerous instances with most signal success. Dr Hill used it in practice for more than twenty-five years. Dr Thornton was quite eminent for his successful application of it." Dr Birch further states, that he had the pleasure of conversing with several highly educated and sensible persons, who declared that their lives were saved by the inhalation of oxygen, under Dr Thornton's care, thirty years ago; and that Dr Riadore, and a few other gentlemen in the profession, had used oxygen more recently with successful results. The Doctor adds, "We have amply sufficient facts to support the claim I now make to have this agent admitted into general practice."*

Mr Samuel Parkes, the author of the "Chemical Catechism," published in 1815, is a very enthusiastic advocate of the use of oxygen, as a most effectual remedy in many diseases, and peculiarly so in those female complaints which arise from want of sufficient tone and vigour in the system. Mr Parkes was so confident of the value of this remedy, and so sanguine of its results in such cases, that he threw his ideas on the subject into the following verses :—

> "Thus the afflicted, feeble, sickly maid,
> Whose spirits languish, and whose health declines,
> Looks to Pneumatic Chemistry for aid,
> And on that science every hope reclines.
>
> Anxious she marks the mixture of the Airs,
> And joys to hear them bubble and expand,
> While fell Hysteria every sinew tears,
> And overwhelms her with his leaden hand.

* Dr Birch on the Therapeutic Action of Oxygen.

Trembling she grasps the cold metallic tube
  In fear mysterious, and with caution drinks,
While every pulse acquires a kinder throb,
  And still increases as the column sinks.

The heaving lungs inhale the Vital Gas,
  The blood absorbs it as it ebbs and flows;
It gives fresh colour to the fluid mass,
  And the whole frame with pristine vigour glows.

Returning health adorns the roseate cheek,
  And decks the features with its every charm:
While new designs new energies bespeak,
  And Beauty's self resumes her native form.

With emulation now her bosom burns:
  The grateful female cultivates her mind,
And feels the sweetest pleasure while she learns
  The science sent by heaven to bless mankind."

I have for many years entertained a very favourable opinion of the value of oxygen gas as a curative agent adapted for many of those truly distressing maladies which do not admit of even amelioration from any known system of medicine. I refer chiefly to cases of inveterate skin disease, indolent ulcers of long standing, their virulent and intractable character doubtless depending upon a deficient supply of oxygen, and consequent imperfect oxygenation of the blood. For consumption also, and other affections of the chest, complicated with debility and sluggish action of the entire system, and for a variety of others of a similar character, I believe the inhalation of oxygen gas will be found a most valuable and important remedy. Under this conviction I have fitted up an oxygen apparatus at Sciennes Hill Baths. No expense has been spared in getting it into proper working order. There is a furnace and purificator in connection with a gasometer capable of holding ten cubic feet of pure oxygen gas. There is also a smaller gasometer, invented by Mr Barth of London, fitted with a graduated rod, by which the quantity of gas to be administered can be exactly measured. The patients inhale from this instrument through an elastic tube, having a glass mouth-piece.

The erection being so recent, I am not in a position to testify as to its curative action from actual experiment, although there are some cases which have received much benefit from its use. One lady from the west country, who had been suffering for a number of years from irritation of the entire mucous membrane, which had resisted all kinds of treatment, was sent by her medical adviser expressly for the inhalation of oxygen. She had from six to ten pints daily for three weeks, and expressed herself as greatly relieved and improved. There was another case of a young man, who had been afflicted for some years with a most inveterate eczematous belt, extending from the lower tip of one ear to that of the other, running under the chin and over the fore part of the throat. Every remedy that could be thought of had been tried without avail. Even the baths at Harrogate and Aix-la-Chapelle were equally unavailing. This person inhaled eight pints of oxygen gas daily for six weeks, by which, together with the Vapour and Alkaline Baths, he was greatly benefited. The eruption was almost entirely removed, and the health and strength immensely invigorated, when the treatment was suspended in consequence of urgent business in England calling the patient away. I have no hesitation in saying that, had he been able to continue a short time longer, the cure would have been most complete. I might mention some other instances where decided benefit was obtained, although the gas was inhaled only for a very short time. What I have already experienced of the efficacy of this remedy, warrants me in anticipating still greater results, in the treatment of those diseases for which it is suited.

Before bringing this volume to a close, I may just add, that Professor Henderson has had the goodness to send me the following extract of a letter from a lady who formerly received treatment at the Sciennes Hill Baths. The letter is dated from the Baths of Ems, Germany:—" I think the Sciennes Hill Baths may stand competition with those of Germany; and, as far as comfort and cleanliness are concerned, are very superior to the latter. Dr Lawrie should call his Acid Bath 'Kessel Water;' it seems to me quite the same."

THE END.

www.ingramcontent.com/pod-product-compliance
Lightning Source LLC
Chambersburg PA
CBHW022117230426
43672CB00008B/1414